Endodontics in Practice

Endodontics in Practice

C. J. R. Stock, MSc, BDS

C. F. Nehammer, BDS, LDS

1990

B·D·J

Published by the British Dental Association
64 Wimpole Street, London W1M 8AL

Second Edition 1990
Reprinted 1992
Reprinted 1994
Reprinted 1996
Reprinted 1998

ISBN 0 904588 29 7
Typeset by Latimer Trend and Company, Great Britain

Printed in England by Thanet Press Limited, Margate, Kent.

Contents

Preface		vi
Acknowledgements		vii
1	The modern concept of endodontics *C. J. R. Stock*	1
2	Diagnosis and treatment planning *C. J. R. Stock*	3
3	Basic instruments and materials for endodontics *C. J. R. Stock*	11
4	Treatment of endodontic emergencies *C. F. Nehammer*	23
5	Morphology of the root canal system *C. F. Nehammer*	29
6	Preparing the root canal *C. J. R. Stock*	37
7	Filling of the root canal system *C. F. Nehammer*	45
8	Calcium hydroxide, root resorption, perio-endo lesions *C. J. R. Stock*	53
9	Endodontic treatment for children *J. R. Goodman*	69
10	Surgical endodontics *C. F. Nehammer*	75
11	Endodontic problems *C. J. R. Stock*	87
Index		95

Preface

This second edition of *Endodontics in Practice* has been extensively updated and expanded, in keeping with the needs of general practitioners and dental students for higher standards and information on more advanced techniques. The demands of patients are now directed far more frequently towards conserving teeth rather than extraction and prosthetic replacement, so that endodontics is becoming a part of the day-to-day routine. This requires the operator to have a far greater knowledge and expertise than was necessary only a few years ago. Reroot treatment, treating the elderly tooth, removing fractured instruments, mending perforations, and producing apical closure with calcium hydroxide, are techniques that every practitioner is expected to carry out. If the dentist does not have the necessary skills, patients will seek help elsewhere.

This book is designed to guide and instruct both the general practitioner and the dental undergraduate through the current techniques used in practice. Tedious research and lengthy explanations have been omitted in favour of the practical approach. After all, it is no good just talking about it to patients, you have to be able to do it.

Acknowledgements

The Authors wish to thank Mr D. J. Setchell and the staff of the Department of Conservative Dentistry and Professor G. B. Winter of the Department of Children's Dentistry at the Eastman Dental Hospital for their advice and help in providing clinical material. Thanks are also due to Mr W. J. Morgan and the Photographic Department and Mr Andrew Johnson, Professional Photographer Graphic Box, for the illustrations, Mrs Angela Christie for the line drawings, and Mrs Bernadette Restall for the preparation of the typescripts. Finally the authors are grateful to Cottrell and Co for supplying a variety of endodontic armamentaria for the illustrations.

1

The Modern Concept of Endodontics

During the last three decades, research in the field of endodontics has modified the approach to treatment. Lesions of endodontic origin which appear radiographically as areas of radiolucency around the apices or lateral aspects of the roots of teeth are, in the majority of cases, sterile.[1] The areas are caused by toxins produced by microorganisms lying within the root canal system. This finding suggests that the removal of microorganisms from the root canal followed by root filling is the first treatment of choice, and that apicectomy with a retrograde filling can only be second best.[2] Apicectomy with a retrograde filling at the apex is carried out in the hope of merely incarcerating microorganisms within the tooth, but does not take into account the fact that approximately 50% of teeth have at least one lateral canal. The long-term success rate of apicectomy must inevitably be lower than orthograde root treatment.

Research into morphology of the pulp has shown the wide variety of shapes, and the occurrence of two or even three canals in a single root.[3] There is a high incidence of fins which run longitudinally within the wall of the canal and a network of communications between canals lying within the same root. The many nooks and crannies within the root canal system make it impossible for any known technique, either chemical or mechanical, to render it sterile. Strong intracanal medicaments such as paraformaldehyde will not only fail to produce sterilisation but may percolate into the periradicular tissues and damage vital healthy tissue, thus delaying healing. The current feeling is to rely on mechanical cleaning of the canal alone, or the use of mild medicaments which are bactericidal but which do not damage tissue.

Other areas of research have had the significant effect of changing the approach to endodontic treatment. The hollow tube theory put forward by Rickert and Dixon in 1931[4] postulated that tissue fluids entering the root canal stagnated and formed toxic breakdown products which then passed out into the periapical tissues. This theory, that dead spaces within the body must be obturated, formed the basis for filling root canals. However, more recently, a variety of different studies[5–9] have demonstrated that, on the contrary, hollow tubes are tolerated by the body. As a result of this work there are currently two indications for filling a root canal: first, to prevent microorganisms from entering the canal system from the oral cavity or via the blood stream (anachoresis), and, secondly, to stop the ingress of tissue fluid which would provide a culture medium for any residual bacteria within the tooth.

All root canal sealers are soluble and their only function is to fill the minute spaces between the wall of the root canal and the root filling material. Their importance, judged by the number of products advertised in the dental press, has been overemphasised. Despite much research, gutta percha remains the root filling of choice, although it is recognised that a biologically inert, insoluble and injectable paste would be better suited for obturation of the root canal. Most of the new root canal filling techniques are concerned with methods of heating gutta-percha, which

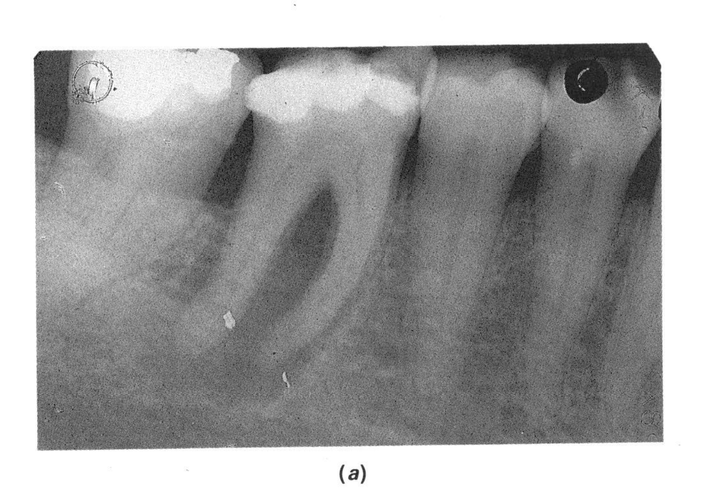

(*a*)

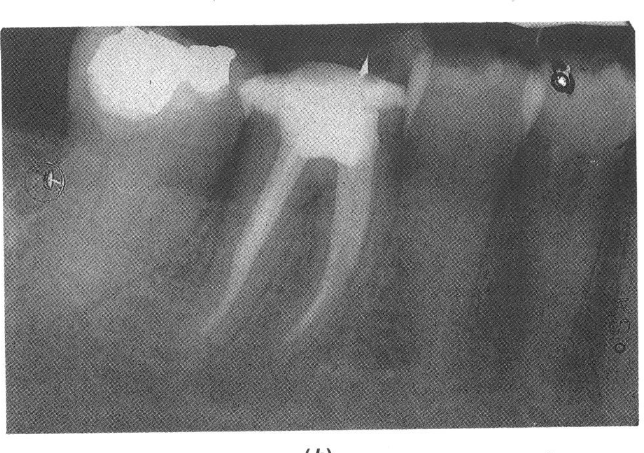

(*b*)

Fig. 1 (*a*) The pre-operative radiograph shows radiolucent areas associated with both root apices and the furcation area of the ⌈6 (36). Root canal treatment was carried out. (*b*) A follow-up radiograph taken 12 months later shows evidence of bony repair with a return to a normal periodontal ligament space around both apices and in the furcation.

makes it softer and easier to adapt to the irregular shape of the canal wall. In summary, the principles of modern endodontic treatment are:

Clean: Remove microorganisms and pulpal debris from the root canal system.
Shape: Produce a gradual smooth taper in the root canal with the widest part coronally and the narrowest part 1 mm short of the apex.
Fill: Obturate the canal system with an inert, insoluble filling material (fig. 1).

References

1 Grossman L I. Bacteriologic status of periapical tissue in 150 cases of infected pulpless teeth. *J Endod* (Special issue) 1982; **8:** 513–515.
2 Harty F J. *Endodontics in clinical practice.* 2nd ed. p 176. Bristol: Wright, 1982.
3 Vertucci F J. Root canal anatomy of the human permanent teeth. *Oral Surg* 1984; **58:** 589–598.
4 Rickert U G, Dixon C M. The controlling of root surgery. In: *Transactions of the Eighth International Dental Congress.* Section 111a. p 15. Paris, 1931.
5 Dubrow H. A method of treating non vital teeth with radiolucent periapical areas. *NY State Dent J* 1964; **30:** 155–159.
6 Goldman M, Pearson A. A preliminary investigation of the 'hollow tube' theory in endodontics: Studies with neotetrazolium. *J Oral Therap Pharmacol* 1965; **1:** 618–626.
7 Torneck C D. Reaction of rat connective tissue to polyethylene tube implants. Part 1: *Oral Surg* 1966; **21:** 379–387. Part 2: *Oral Surg* 1967; **24:** 674–683.
8 Friend L A, Browne R M. Tissue reactions to some root filling materials. *Br Dent J* 1968; **125:** 291–298.
9 Moller A J R *et al.* Influence on periapical tissues of indigenous oral bacteria and necrotic pulp tissue in monkeys. *Scand J Dent Res* 1981; **89:** 475–484.

Further reading

Seidler B. Irrationalised endodontics: N_2 and us too. *J Am Dent Assoc* 1974; **89:** 1318–1331.

2

Diagnosis and Treatment Planning

The importance of correct diagnosis and treatment must be stressed. There are many causes of facial pain and the differential diagnosis can be both difficult and demanding. All the relevant information must be collected; this includes a case history and the results of both a clinical examination and diagnostic tests. Only at this stage can the cause of the problem be determined and the treatment for the patient planned.

Case history

The purpose of a case history is to discover whether the patient has any general or local condition that might alter the normal course of treatment (Table I). In addition, a description of the patient's symptoms in his or her own words and a history of relevant dental treatment should be noted. An example of the particulars required on a patient's folder is illustrated.

Medical history

There are no medical conditions which specifically contra-indicate endodontic treatment, but there are several which require special care (Table II). If there is any doubt about the state of health of a patient, his/her general medical practitioner should be consulted before any endodontic treatment is commenced. This also applies if the patient is on medication, such as corticosteroids or an anticoagulant. Antibiotic cover is recommended for certain medical conditions, depending upon the complexity of the procedure and the degree of bacteraemia expected, but the type of antibiotic and the dosage are under continual review and dental practitioners should be aware of current views. Some guidelines are given in Table III. Patients who are predisposed to endocarditis and are given a prophylactic antibiotic should be advised to report even a minor febrile illness which occurs up to 2 months following endodontic treatment. Prior to endodontic surgery, it is useful to prescribe aqueous chlorhexidine (0·2%) mouthwash.

Patient's complaints

Listening carefully to the patient's description of his/her symptoms can provide invaluable information. It is quicker and more efficient to ask patients specific, but not leading, questions about their pain. Examples of the type of questions which may be asked are given below.

(1) How long have you had the pain?
(2) Do you know which tooth it is?
(3) What initiates the pain?
(4) How would you describe the pain?
  Sharp or dull
  Throbbing
  Mild or severe
  Localised or radiating
(5) How long does the pain last?
(6) When does it hurt most? During the day or at night?
(7) Does anything relieve the pain?

It is usually possible to decide, as a result of questioning the patient, whether the pain is of pulpal, periapical or periodontal origin, or, that it is non-dental in origin. It is not possible to diagnose the histological state of the pulp

Table I Example of medical history questions for a patient's folder

Medical history		
Rheumatic fever	Yes	No
If yes, is there any cardiac damage	Yes	No
Hypertension or cardiac disease	Yes	No
Allergies	Yes	No
Hepatitis	Yes	No
Pregnant	Yes	No
Upper resp. tract infections	Yes	No
Taking any drugs now	Yes	No
Anticoagulants	Yes	No
Steroids	Yes	No
Insulin	Yes	No
Tranquillisers	Yes	No
Other	Yes	No
Under treatment by GP or hospital	Yes	No
Serious illness in the past 3 years	Yes	No
Further medical history		

Table II Some medical conditions relevant to endodontic treatment and the precautions which should be taken by the dentist

Medical condition	Precautions taken
History of infective endocarditis	Regarded as special high risk group. Patients usually referred to hospital for treatment. Antibiotic cover required.
Congenital cardiac abnormality	Consider antibiotic cover (Table III)
Rheumatic fever or Sydenham's chorea	Consider antibiotic cover.
Artificial heart valves	Patient's cardiologist to advise antibiotic cover.
Prosthesis for total replacement of a joint	Consider antibiotic cover (advice from orthopaedist varies)
Cardiovascular disease	GMP to advise alteration of drug therapy. Non-surgical endodontics preferred. Analgesics to reduce post-operative pain. Appointments not to exceed 1 hour.
Hypertension	GMP to advise alteration of drug therapy. Non-surgical endodontics preferred. Analgesics to reduce post-operative pain. Appointments not to exceed 1 hour.
Blood disease (haemophilia)	Root canal treatment preferred to extraction. No local anaesthetic if possible. Use of devitalising pastes. Care taken not to lacerate gingivae.
Patients on anticoagulants	GMP to advise.
Patients on corticosteroids (currently or during past 12 months)	200 mg oral 2 hours pre-operatively or double dose night before and on day of operation.
Diabetes	No general anaesthetic if possible. Glucose kept in surgery. Healing will be retarded. Antibiotic cover if surgery intended or infection present. GMP to advise any alteration of patient's drug therapy.
Hepatitis	Dangers: (1) Operator contracting disease. (2) Cross-infection via contaminated instruments. GMP to check if patient is carrier. Treat with caution: (1) Rubber gloves, mask and glasses. (2) Low-speed instruments. (3) Dispose of instruments which may pick up microorganisms, eg files, burs, etc. (4) Wash down operating area with 2% glutaraldehyde solution and place all instruments, etc, used on patient in the same solution for 1 hour before sterilisation. (5) Treat at end of day.
Chronic renal failure	Take all precautions to prevent hepatitis cross-infection if patient on kidney machine.
Immunosuppressed states: Patients on corticosteroids or drugs to maintain organ transplants	Antibiotic cover if infection present. Possible corticosteroid supplement by GMP during and after endodontic treatment.
Radiotherapy for malignant disease	Any extractions necessary carried out before radiotherapy to prevent intractable radionecrosis. Root canal treatment preferred to extraction after therapy.
Sexually transmitted disease	If HIV-positive but no symptoms treat as for hepatitis. If HIV-positive and with symptoms, refer to specialist unit. Syphilitic lesions may resemble a draining sinus.
Other debilitating diseases, ie asthma, hay fever and skin rashes	Patient could be sensitive to drugs: only prescribe drugs which patient has taken previously.

Table III Antimicrobial prophylaxis against infective endocarditis

The table shows regimes which are advised currently and may be adopted when antibiotic cover is necessary. Advice on particular regimes is changing constantly and the operator would be advised to check with current opinion.

	Not allergic to penicillin No recent penicillin	Allergy to or recent treatment with penicillin	
Local anaesthesia	Amoxycillin oral 3 g 1 hour pre-op; children under 10 one half adult dose. One dose dispersible tablets preferred for diabetics.	Erythromycin oral 1·5 g, then 0·5 g orally 6 hours later. Children under 10 one half adult dose.	
General anaesthesia	Amoxycillin oral 3 g 4 hours pre-op	Penicillin within previous 2 weeks	Allergic to penicillin
	or IM 1 g 1 hour pre-op. and 0·5 g orally 6 hours later.	Amoxycillin 1 g IM plus gentamycin 120 mg IM followed by amoxycillin 0·5 g orally 6 hours later.	Vancomycin 1 g IV plus gentamycin 120 mg IV

from the clinical signs and symptoms. In cases of pulpitis, the decision the operator must make is whether the pulpal inflammation is reversible, in which case it may be treated, or irreversible, in which case the pulp or tooth must be removed.

In early pulpitis the patient cannot localise the pain to a particular tooth or jaw because the pulp does not contain any proprioceptive nerve endings. As the disease advances and the periapical region becomes involved, the tooth will become tender and the proprioceptive nerve endings in the periodontal ligament are stimulated.

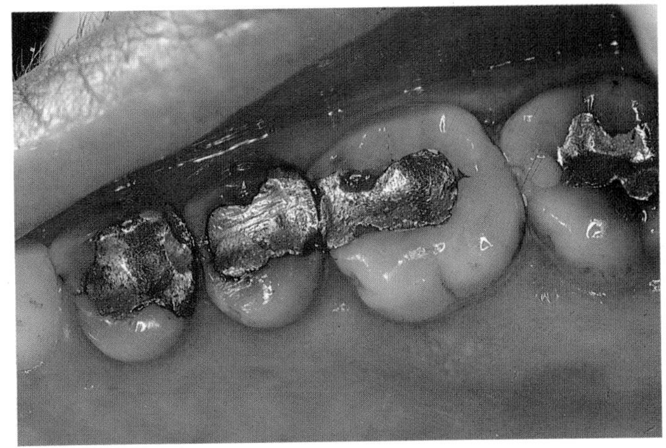

Fig. 1 An assessment is made of the patient's general dental condition.

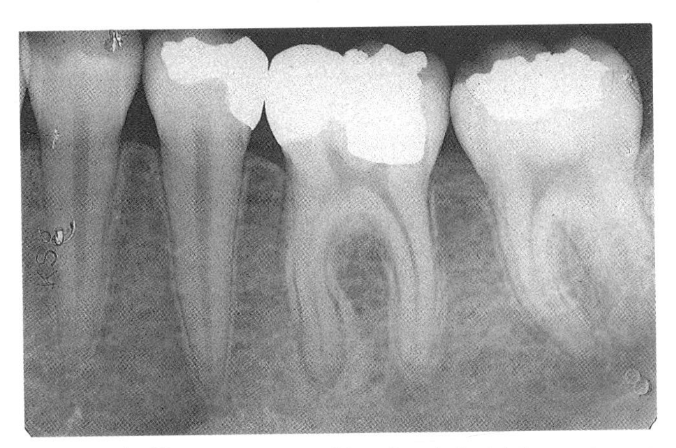

Fig. 2 The |6 (36) shows a widened periodontal ligament space around the apex of the mesial root with changes in the surrounding bony trabeculae.

Clinical examination

A clinical examination of the patient is carried out after the case history has been completed. The temptation to start treatment on a tooth without examining the remaining dentition must be resisted. Problems must not be dealt with in isolation and any treatment plan should take the entire mouth and the patient's general medical condition and attitude into consideration.

Extra-oral examination

The patient's face and neck are examined and any swelling, tender areas, lymphadenopathy, or extra-oral sinuses noted.

Intra-oral examination

An assessment of the patient's general dental state is made, noting in particular the following aspects (fig. 1):

(1) Standard of oral hygiene.
(2) Amount and quality of restorative work.
(3) Prevalence of caries.
(4) Missing and unopposed teeth.
(5) General periodontal condition.
(6) Presence of soft or hard swellings.
(7) Presence of any sinus tracts.
(8) Discoloured teeth.
(9) Tooth wear and facets.

Diagnostic tests

Most of the diagnostic tests used to assess the state of the pulp and periapical tissues are relatively crude and unreliable. No single test, however positive the result, is sufficient to make a firm diagnosis of reversible or irreversible pulpitis. There is a general rule that before drilling into a pulp chamber there should be two independent positive diagnostic tests. An example would be a tooth non-vital to the electric pulp tester and tender to percussion.

Palpation

The tissues overlying the apices of any suspect teeth are palpated to locate tender areas. The site and size of any soft or hard swellings are noted and examined for fluctuation and crepitus.

Percussion

Gentle tapping with a finger both laterally and vertically on a tooth is sufficient to elicit any tenderness. It is not necessary to strike the tooth with a mirror handle, as this invites a false positive reaction from the patient.

Mobility

The mobility of a tooth is tested by placing a finger on either side of the crown and pushing with one finger while assessing any movement with the other. Mobility may be graded as: 1—slight (normal), 2—moderate, and 3—extensive movement in a lateral or mesiodistal direction combined with a vertical displacement in the alveolus.

Radiography

In all endodontic cases, a good intra-oral parallel radiograph of the root and periapical region is mandatory. Radiography is the most reliable of all the diagnostic tests and provides the most valuable information. A routine radiograph may be the first indication of the presence of pathology (fig. 2). The disadvantage of radiography is that the early stages of pulpitis are not normally evident on the radiograph. If a sinus is present and patent, a small sized gutta-percha point should be inserted and teased, by rolling gently between the fingers, as far along the sinus tract as possible. A radiograph taken with the gutta-percha point in place will often show the cause of the problem (fig. 3).

Pulp testing

The electric pulp tester is an instrument which uses gradations of electric current to excite a response from the nervous tissue within the pulp. Both alternating and direct

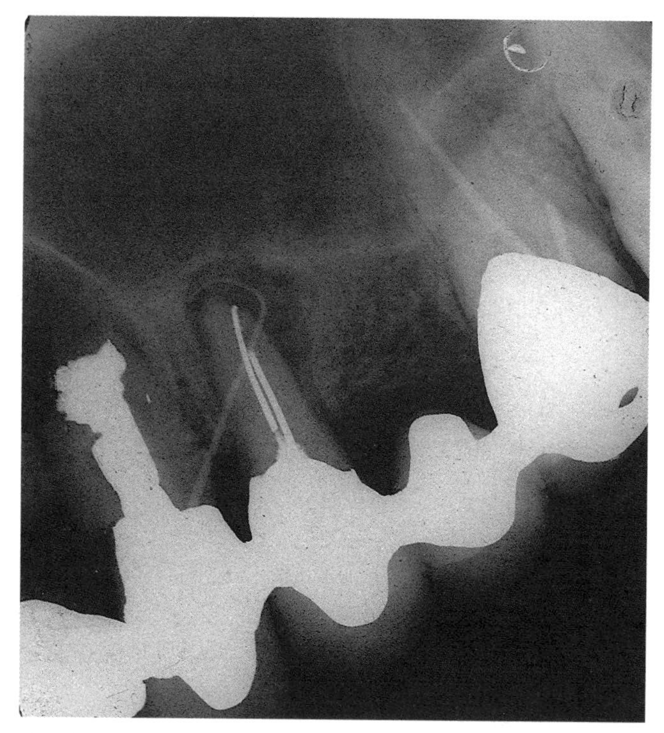

Fig. 3 A gutta-percha point placed in the sinus opposite 4|(14) which is an abutment for a bridge. The gutta-percha points to a periapical area.

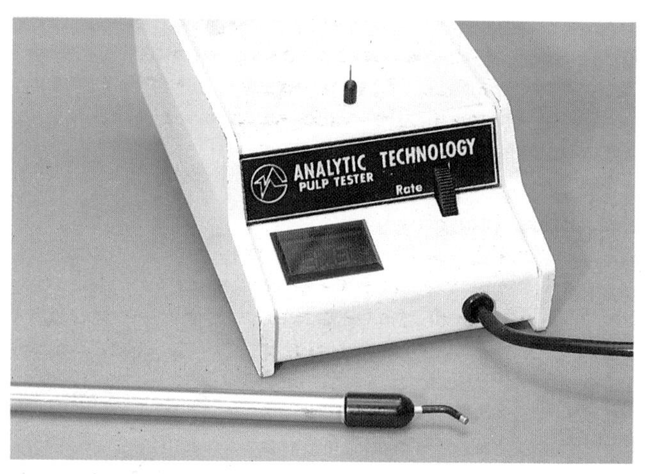

Fig. 4 Pulp tester (Analytic Technology, Redmond, WA 98052, USA) with digital display 0–80. It switches on automatically when the probe tip touches the tooth and turns off when tooth contact is broken after a 15-second delay. The only control of the EPT is the rate of increase of the electrical stimulus.

current pulp testers are available, although there is little difference between them. Most pulp testers manufactured today are monopolar (fig. 4).

Pulp testers should only be used to assess vital or non-vital pulps; they do not quantify disease, nor do they measure health and should not be used to judge the degree of pulpal disease. There are several disadvantages to electric pulp testing. No indication is given of the state of the vascular supply which would indicate more accurately the degree of pulp vitality, false positive readings occur due to stimulation of nerve fibres in the periodontium, and, finally, posterior teeth may give misleading readings since a combination of vital and non-vital root canal pulps may be present.

The use of gloves in the treatment of all dental patients has produced problems with electric pulp testing. There is a lip electrode attachment available which may be used, but a far simpler method is to ask the patient to hold on to the metal handle of the pulp tester. The patient is asked to let go of the handle if they feel a sensation in the tooth being tested.

Doubt has been cast on the efficacy of pulp testing the corresponding tooth on the other side of the mid-line for comparison, and it is suggested that only the suspect teeth are tested. The teeth to be tested are dried and isolated with cotton wool rolls. A conducting medium should be used; the one most readily available is toothpaste. Pulp testers should not be used on patients with pacemakers because of the possibility of electrical interference.

Teeth with full crowns present problems with pulp testing. A pulp tester is available (fig. 5) with a special point fitting which may be placed between the crown and the gingival margin. There is little to commend the technique of cutting a window in the crown in order to pulp test.

Thermal pulp testing

This involves applying either heat or cold to a tooth, but neither test is particularly reliable and may produce either false positive or false negative results.

Heat

There are several different methods of applying heat to a tooth. The tip of a gutta-percha stick may be heated in a

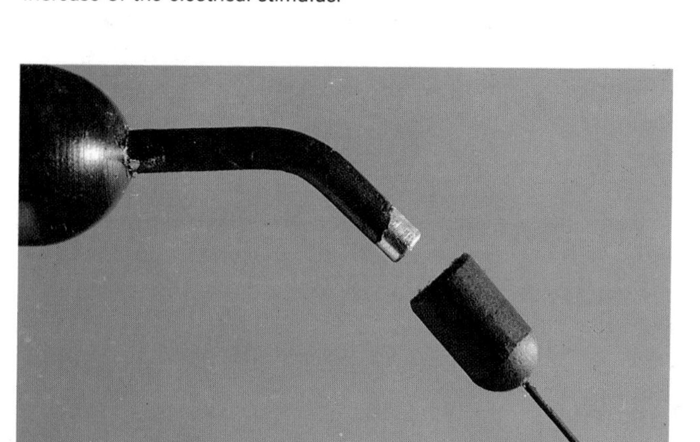

(*a*)

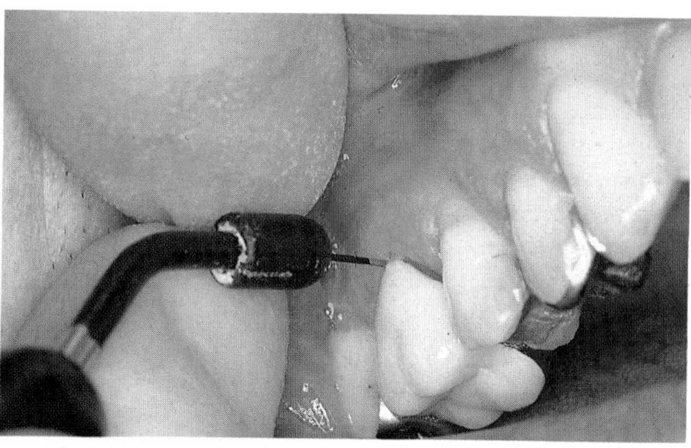

(*b*)

Fig. 5 A special probe tip may be fitted to this pulp tester (*a*) which makes it possible to test a crowned tooth providing contact can be made with a small area of tooth (*b*).

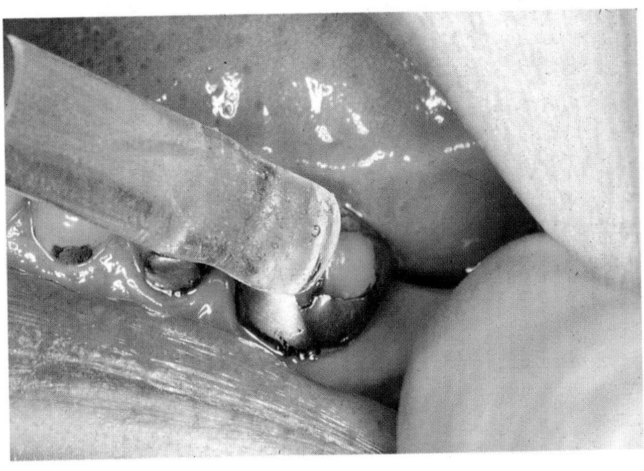
(a)

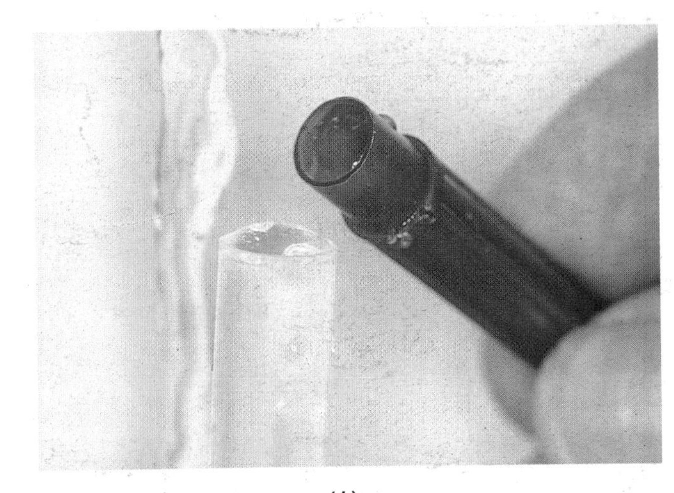
(b)

Fig. 6 (*a*) An ice stick applied to a tooth is a simple test. (*b*) Plastic covers for a hypodermic needle are filled with water.

flame and applied to a tooth. It is advisable to coat the tooth with vaseline to prevent the gutta-percha sticking and causing unnecessary pain to the patient. Another method is to use the heat generated from a rubber wheel in a standard handpiece.

Cold

An ethyl chloride spray on a pledget of cotton wool or an ice stick may be applied to the suspect tooth. Ice sticks are made by filling the plastic covers from a hypodermic needle with water and placing in a refrigerator. When required for use one cover is warmed and removed to provide the ice stick (fig. 6).

Local anaesthetic

In cases where the patient cannot locate the pain and the thermal test is negative, a reaction may be obtained by asking the patient to sip hot water from a cup. The patient is instructed to hold the water first against the mandibular teeth on one side and then by tilting the head, to include the maxillary teeth. If a reaction occurs, an intraligamental injection may be given to anaesthetise the suspect tooth and hot water is then again applied to the area; if there is no reaction, the pulpitic tooth has been identified. It should be borne in mind that a better term for intraligamental is intraosseous, as the local anaesthetic will pass into the medullary spaces round the tooth and so possibly also affect the proximal teeth.

Wooden stick

If a patient complains of pain on chewing and there is no evidence of periapical inflammation, an incomplete fracture of the tooth may be suspected. Biting on a wood stick in these cases can elicit pain, usually on release of biting pressure.

Fibre-optic light

A powerful light can be used for transilluminating teeth to show interproximal caries, fracture, opacity, or discoloration. To carry out the test, the dental light is turned off and the fibre-optic light placed against the tooth at the gingival margin and the beam directed through the tooth. If the crown of the tooth is fractured, the light will pass through the tooth until it strikes the stain lying in the fracture line; the tooth beyond the fracture will appear darker.

Cutting a test cavity

When other tests have given an indeterminate result, a test cavity may be cut in a tooth which is believed to be pulpless. In the author's opinion, this test can be unreliable as the patient may give a positive response although the pulp is necrotic. This is because nerve tissues can continue to conduct impulses for some time in the absence of a blood supply.

Treatment planning

Having taken the case history and carried out the relevant diagnostic tests, the patient's treatment is then planned. The type of endodontic treatment chosen must take into account the patient's medical condition and general dental state. The indications and contra-indications for endodontics are given below and the problems of reroot treatment discussed. The treatment of fractured instruments, perforations and perio-endo lesions are discussed in subsequent chapters.

Indications for endodontics

All teeth with pulpal or periapical pathology are candidates for endodontics. There are also situations where elective endodontics is the treatment of choice.

Post space

A vital tooth may have insufficient tooth substance to retain a jacket crown so the tooth may be root-treated and a post crown fitted (fig. 7).

Overdenture

Decoronated teeth retained in the arch to preserve alveolar bone must be root-treated.

Teeth with doubtful pulps

Root treatment should be considered for any tooth with doubtful vitality (fig. 8) if it requires an extensive restoration, particularly if it is to be a bridge abutment.

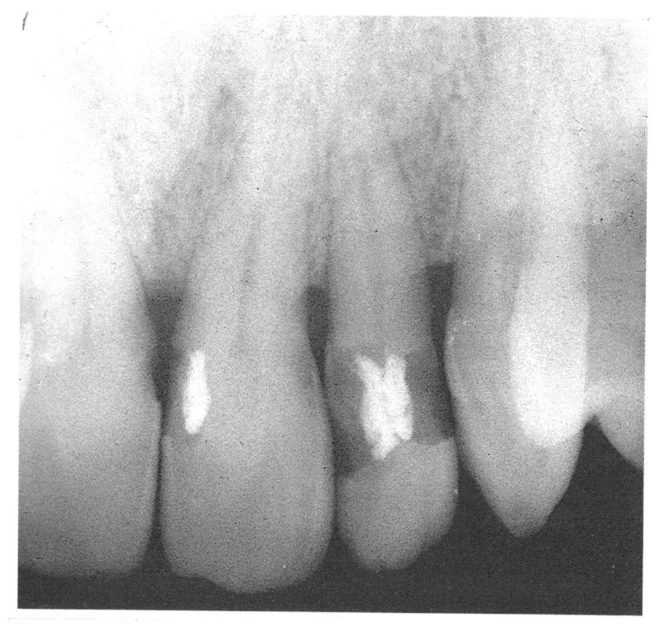

Fig. 7 The ⌊2 (22) requires a crown but will need elective devitalisation and root treatment to provide post space for a post crown.

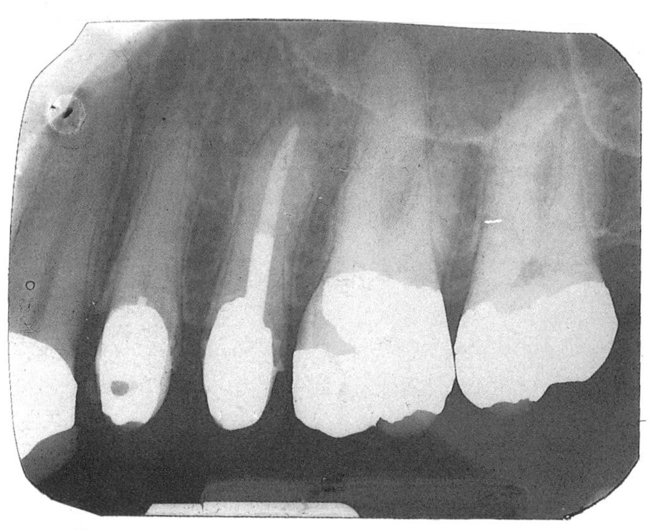

Fig. 8 The ⌊6 (26) was to be crowned but an amalgam core had been placed without a local anaesthetic and the patient had felt no pain. The vitality of the ⌊6 (26) was therefore in doubt and the tooth should be root-filled before crowning.

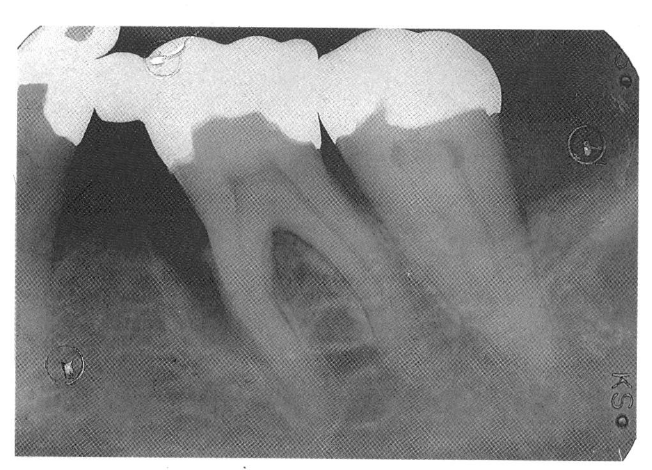

Fig. 9 There is deep pocketing associated with the mesial root of the ⌈6 (36). If this cannot be made healthy with periodontal treatment, the tooth may be root-treated and the mesial root resected.

Risk of exposure

Preparing teeth for crowning in order to align them in the dental arch can risk traumatic exposure. In some cases these teeth should be electively root-treated.

Periodontal disease

In multi-rooted teeth there may be deep pocketing associated with one root or the furcation (fig. 9). The possibility of elective devitalisation following the resection of a root should be considered (*see* chapter 9).

Pulpal sclerosis following trauma

Review periapical radiographs should be taken following trauma. If progressive narrowing of the pulp space is seen due to secondary dentine, elective endodontics may be considered while the coronal portion of the root canal is still patent. This may occasionally apply after a pulpotomy has been carried out.

Contra-indications to endodontics

The medical conditions which require special precautions prior to endodontic treatment have already been listed. There are, however, other conditions both general and local, which may contra-indicate endodontics.

General

Inadequate access

A patient with restricted opening or a small mouth may not allow sufficient access for endodontic treatment. A rough guide is that it must be possible to place two fingers between the mandibular and maxillary incisor teeth so that there is good visual access to the areas to be treated. An assessment for posterior endodontic surgery may be made by retracting the cheek with a finger. If the operation site can be seen directly with ease, then the access is sufficient.

Poor oral hygiene

Endodontics should not be carried out unless the patient is able to maintain his/her mouth in a healthy state or can be taught to do so.

Patient's general medical condition

The patient's physical or mental condition due to, for example, a chronic debilitating disease, or old age, may preclude endodontic treatment. Similarly, the patient at high risk to infective endocarditis, for example one who has had a previous attack, may not be considered suitable for complex endodontic therapy.

Patient's attitude

Unless the patient is sufficiently well motivated, a simpler form of treatment is advised.

Local

Tooth not restorable

It must be possible, following endodontic treatment, to restore the tooth to health and function (fig. 10). The finishing line of the restoration must be supracrestal and preferably supragingival.

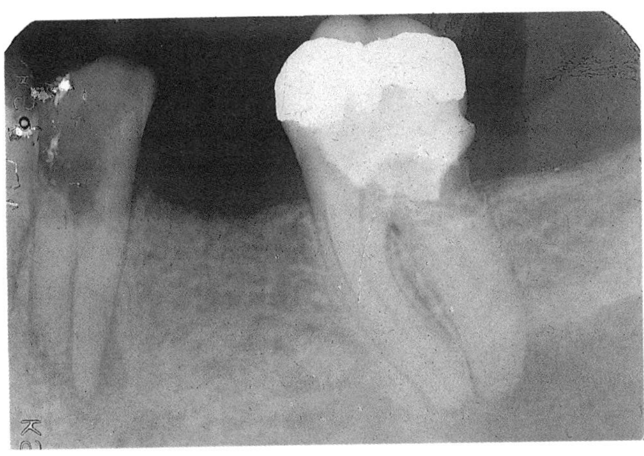

Fig. 10 The |7 (37) shows loss of tooth substance of the distal root below the bony crest. This tooth would be almost impossible to restore and should be extracted.

Insufficient periodontal support
Provided the tooth is functional and the attachment apparatus healthy, or can be made so, endodontic treatment may be carried out.

Non-strategic tooth
Extraction should be considered rather than endodontic treatment for unopposed and non-functional teeth.

Root fractures
Incomplete fractures of the root have a poor prognosis if the fracture line communicates with the oral cavity as it becomes infected. For this reason, vertical fractures will often require extraction of the tooth while horizontal root fractures have a more favourable prognosis (fig. 11).

Massive internal or external resorption
Both types of resorption may eventually lead to pathological fracture of the tooth. Internal resorption ceases immediately the pulp is removed and, provided the tooth is sufficiently strong, it may be retained. Most forms of external resorption will continue (*see* Chapter 9) unless the defect can be repaired and made supragingival, or arrested with calcium hydroxide therapy.

Bizarre anatomy
Exceptionally curved roots (fig. 12), dilacerated teeth, and congenital palatal grooves may all present considerable difficulties if root canal treatment is attempted. In addition, any unusual anatomical features related to the roots of the teeth should be noted as these may affect prognosis.

Reroot treatment

One problem which confronts the general dental practitioner is to decide whether an inadequate root treatment requires replacement. The questions the operator should consider are given below (fig. 13).

(1) Is there any evidence that the old root filling has failed?
 (a) Symptoms from the tooth.
 (b) Radiolucent area is still present or has increased in size.
 (c) Presence of sinus tract.

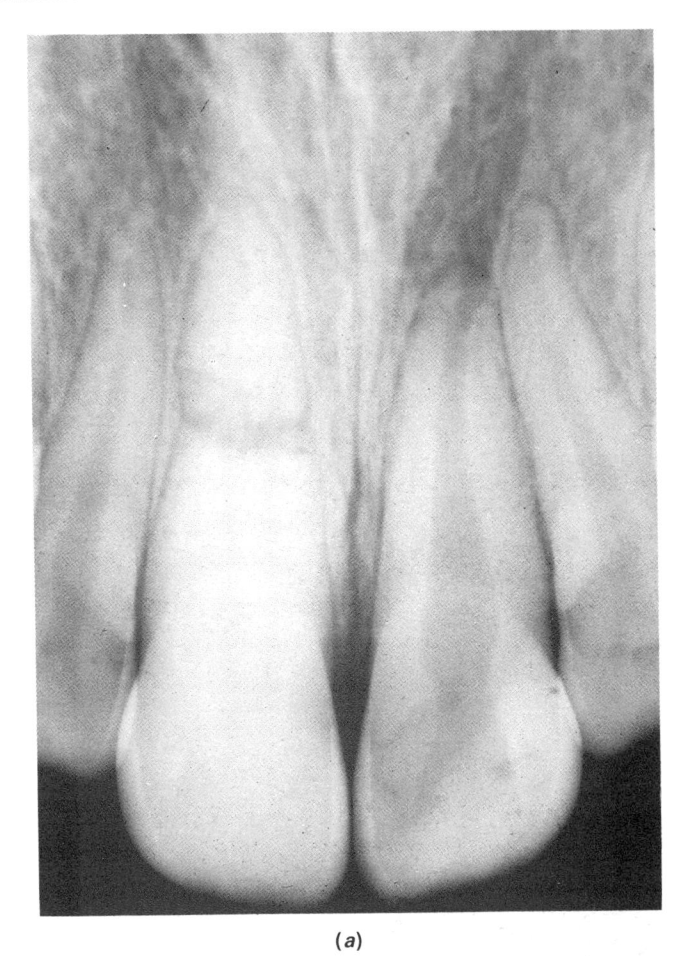

(*a*)

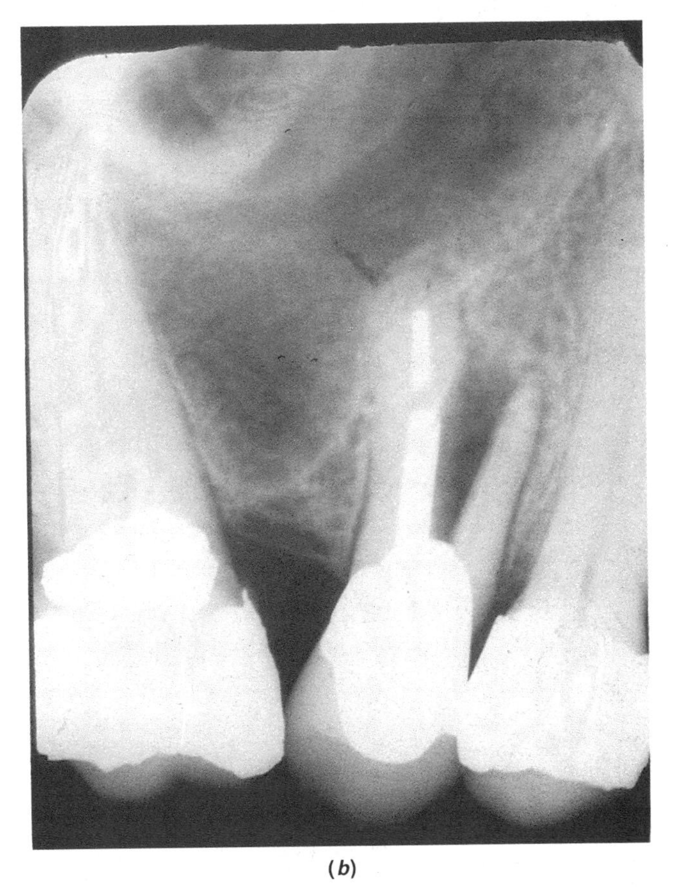

(*b*)

Fig. 11 (*a*) Horizontal fracture of 1|(11) due to trauma 13 years previously. No treatment was carried out. The tooth remains symptomless and firm although there is some loss of translucency. (*b*) Vertical root fracture; the tooth was extracted (*continued on p. 10*).

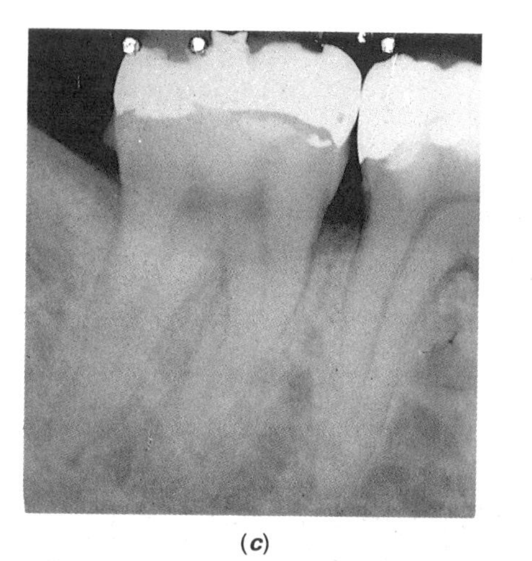

(*c*)

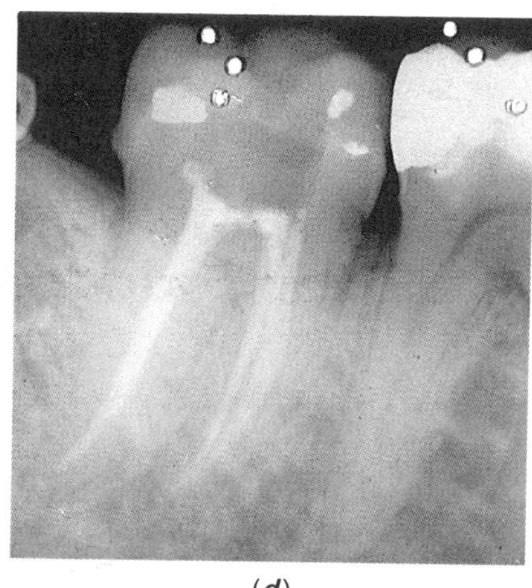

(*d*)

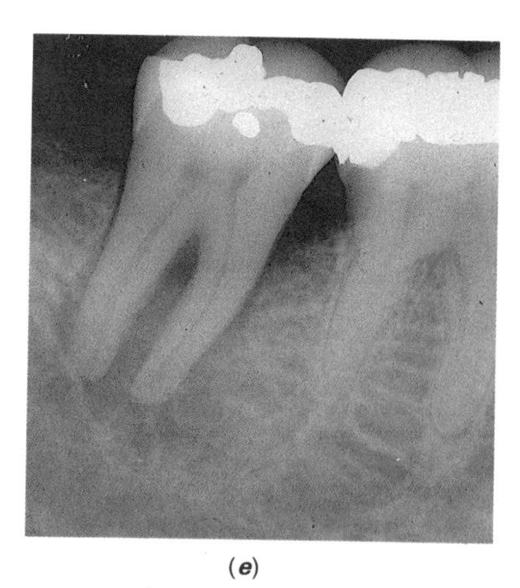

(*e*)

Fig. 11 (*continued*) (*c*) The 7| (47) had a fine fracture line involving the distal wall of the distal root. The fracture did not extend into the floor of the pulp chamber. (*d*) The tooth was root-filled and had a cusp covered cast restoration placed. (*e*) The |7 (37) had a definite fracture line running mesiodistally and involving the floor of the pulp chamber. The tooth was extracted.

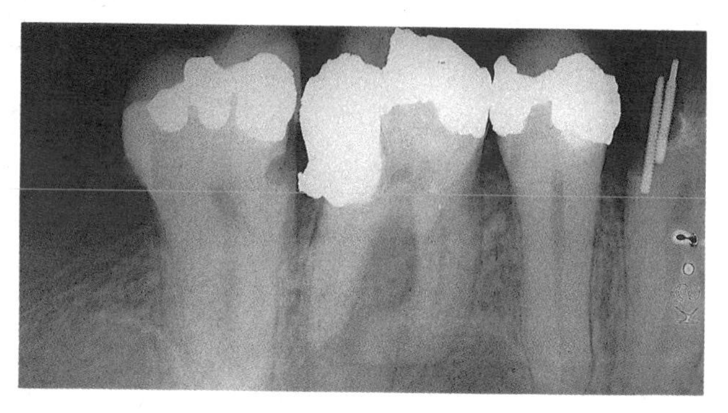

Fig. 12 The |6 (36) has a severely curved mesiobuccal root. Negotiation and preparation of this root canal would be difficult but not impossible.

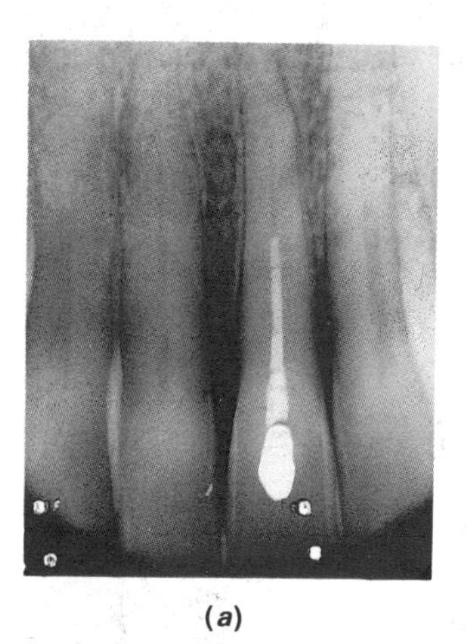

(*a*)

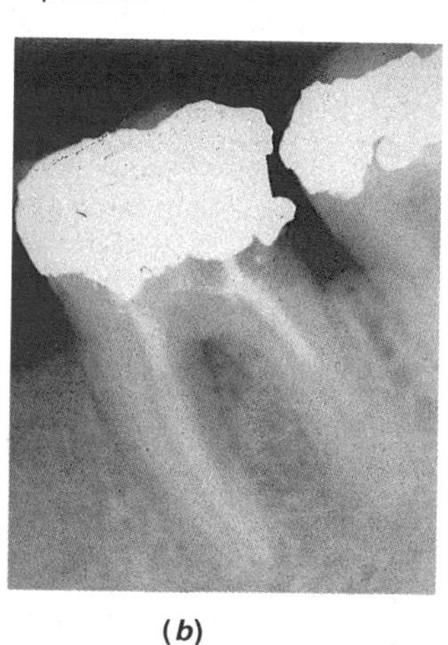

(*b*)

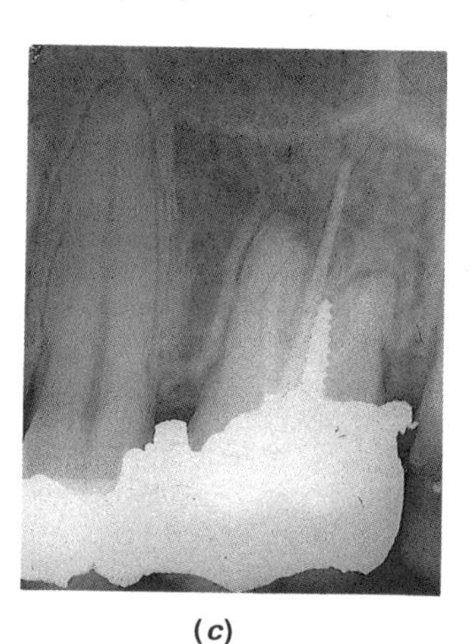

(*c*)

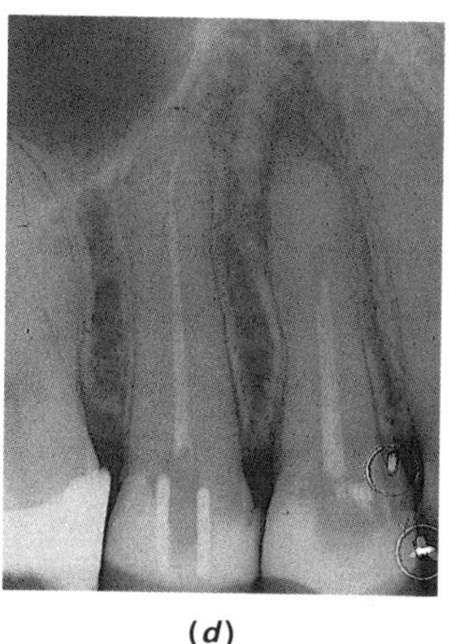

(*d*)

Fig. 13 (*a*) The 1| (11) had been root-treated 4 years previously and had remained symptomless. The coronal restoration was intact. It was decided not to reroot-fill. (*b*) The 6| (46) was symptomless. The obvious caries beneath the distal amalgam meant the restoration required renewing. The inadequate root filling must be removed and an attempt made to reroot-treat. (*c*) The |6 (26) was slightly TTP. The buccal roots are short, probably owing to resorption, and the palatal root is not visible on the radiograph. The retention screw could be in the furcation. The prognosis is poor and the tooth should be extracted. (*d*) The |4 (24) has been inadequately root-treated and there is obvious periapical pathology. The tooth should be reroot-filled. The |5 (25) has also been poorly condensed, but there is no pathology evident and the tooth is symptomless. It should be root-filled if the coronal restoration requires replacing.

(2) Does the crown of the tooth need restoring?
(3) Is there any obvious fault with the present root filling which could lead to failure?

The final decision by the operator on the treatment plan for a patient will be governed by the level of his/her own skill and knowledge. General dental practitioner cannot become experts in all fields of dentistry and should learn to be aware of their own limitations. The treatment plan proposed should be one which the operator is confident he/she can carry out to a high standard.

References

1 Scully C. *Hospital dental surgeons guide.* p 63. London: British Dental Journal, 1985.
2 Scully C. *Hospital dental surgeons guide.* p 64. London: British Dental Journal, 1985.

3

Basic Instruments and Materials for Endodontics

Many dental practitioners find it difficult to resist new gadgets, and there are many made specifically for endodontics. New instruments and materials are frequently sold with the promise of simplifying a technique, shortening the time taken, or even increasing the success rate. Unfortunately, it is rare to find that these promises can be fulfilled, and the result is often surgery cupboards containing unwanted endodontic armamentaria. It would be impossible to cover all the instruments and materials used in endodontics in one chapter, but it is hoped to mention most of the basic equipment and discuss some of the newer items. For continuity, some instruments will be described in the relevant chapters.

Instrument pack

A basic pack of instruments must be available for routine root canal procedures. An example is given in figure 1. The mouth mirror* is front surface reflecting to prevent a double image. Endolocking tweezers allow small items to be gripped safely and passed between nurse and operator. A fine, strong, long probe is required to detect canal orifices. The excavator is long shanked, with a small blade to allow access into the pulp chamber. The pocket-measuring probe should have a narrow shank, a blunt tip and clearly visible gradations. A furcation probe is useful to check for the presence of furcation involvement. Other items which are usually included are a flat plastic, a Briault probe, sterile cotton wool rolls (size 2), sterile cotton wool pledgets, artery forceps to grip a periapical radiograph and, finally, a metal ruler. Paper points are also required, and the simplest method of storage and use is to acquire presterilised packs with five points in each pack.

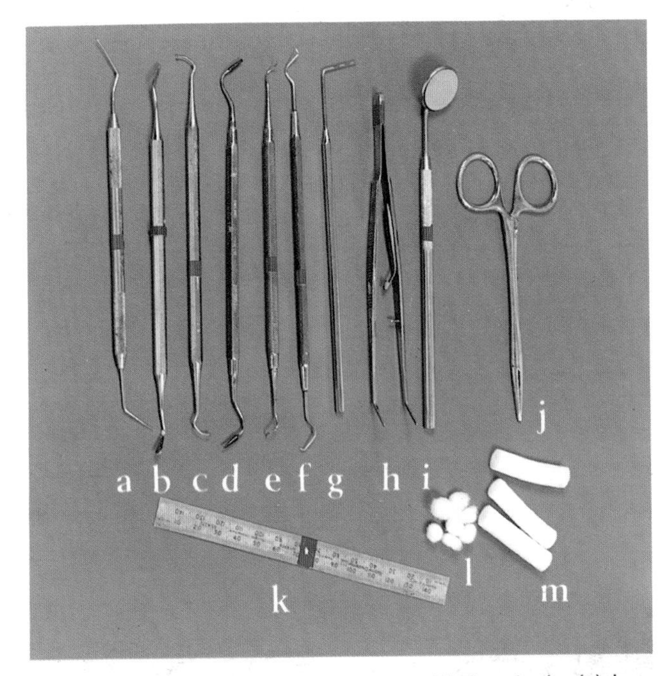

Fig. 1 Instrument pack. (*a*) Canal probe; (*b*) flat plastic; (*c*) long shank excavator; (*d*) Mortenson plugger; (*e*) American pattern probe no. 3; (*f*) Briault probe; (*g*) Marquis perio probe; (*h*) endo locking tweezers; (*i*) front surface reflecting mirror; (*j*) artery forceps; (*k*) metal ruler; (*l*) cotton wool pledgets; (*m*) cotton wool rolls.

Trays

There are two types of tray used in endodontics: a metal tray with a lid which contains a set of sterile instruments and an open tray on which may be laid a sterile pack of instruments. An example of each type is illustrated in figure 2. The autoclave has to be large enough to accept a tray with a lid, and although ideal for a large clinic or hospital, the open tray is more suitable in general practice. The open tray shown, which comes in several colours, is made of plastic, so only cold sterilising solutions may be used. If a tray system is used, it is useful to have a cupboard in the surgery with partitions to store clean and dirty trays (fig. 3). Figure 4 shows the cervical tray over the patient. The tray contains a sterile instrument pack and sterile sponge which is cut to fit one of the compartments.

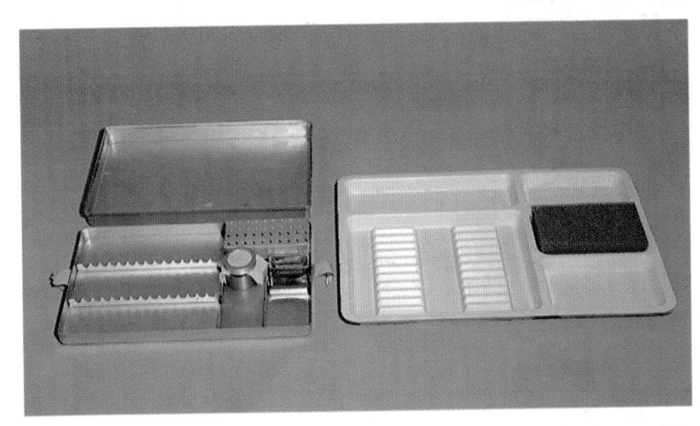

Fig. 2 Trays. On the left, the RAF tray is 286 mm × 185 mm. The inserts include an instrument stand, two medicament trays and a 'clean grip'. On the right, the double plastic tray is 383 mm × 266 mm. An artificial sponge may be cut to fit a compartment and be retained with pins. The sponge is sterilised in the autoclave.

*Details will be found at the end of the chapter

Fig. 3 A cupboard built into the dental surgery, designed for the tray system. Clean trays are stored on the left and dirty trays on the right.

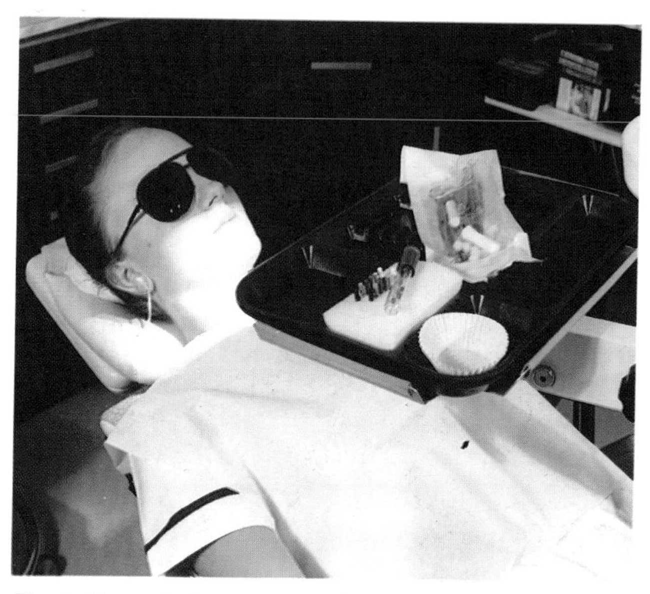

Fig. 4 The cervical tray over the patient contains an instrument pack, a sterile sponge cut to fit a compartment in the tray, and a paper baking case for refuse. The patient is wearing eye protectors and a waterproof bib.

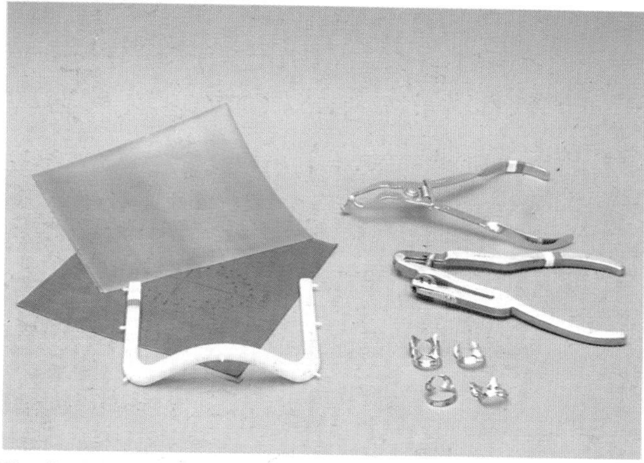

Fig. 5 Rubber dam kit. The rubber dam is extra heavy, and two types are shown here with a frame. Also shown are clamp forceps, a punch and four clamps: Ash 14a, Ivory W8, Ivory W8a and Ivory no. 1.

Patient protection

Glasses are needed to protect the patient's eyes. Figure 4 also shows a waterproof bib being worn, as clothes require protection against accidental spillage of sodium hypochlorite.

Rubber dam

Rubber dam is necessary in root canal treatment for three reasons:

(1) To prevent the patient swallowing or inhaling root canal instruments or medicaments.
(2) To provide a field which is relatively free from oral contamination.
(3) To give good visual access by retracting the lips and tongue.

A basic kit for rubber dam equipment is shown in figure 5. Rubber dam is supplied in several different weights, but only heavy (0·010 inch) or extra heavy (0·012 inch) are recommended because they are less likely to tear. Different colours are available and are a matter of personal preference; black and green are widely used.

A rubber dam frame should be shaped to the face and wide enough to allow good access. Most of the newer designs are made of radiolucent plastic. Forceps and a rubber dam punch are required. The most important aspect of the punch is that it should produce a neat hole in the dam with no tags. A wide variety of clamps are available, although only a few are needed for everyday practice.

Methods of application of rubber dam are described by Reuter.[1] As a general rule in root canal treatment only one tooth need be isolated in the posterior part of the mouth. For the incisors, it is usually simpler to isolate several teeth so that the dam is pulled well away from the cingulum. The easiest method of applying rubber dam is to fit the clamp on to the tooth and then, having punched a hole in the dam, stretch it over the bow and tooth. Rubber dam may be retained on the incisor teeth using wedges cut from the corner of the dam. Clamps may be placed on the premolars over the rubber dam to give better access if needed.

Radiographic equipment

A quick, reliable x-ray film processor is essential for endodontics. A manual processor with rapid developing and fixing solutions can produce a radiograph for viewing in under a minute; an example is shown in figures 6 and 7.

Parallel radiography is another requirement for endodontics, because it gives an undistorted view of the teeth and surrounding structures and is repeatable, thus allowing assessment of periapical healing. A device is required to hold the x-ray film parallel to the tooth. Figure 8 shows an example of a popular holder.

An x-ray viewer and some form of magnification are needed to examine periapical films. It is helpful if glare from the light around the radiograph can be excluded (fig. 9).

Hand instruments

Hand root canal cutting instruments are manufactured to a size and type advised by the International Standards

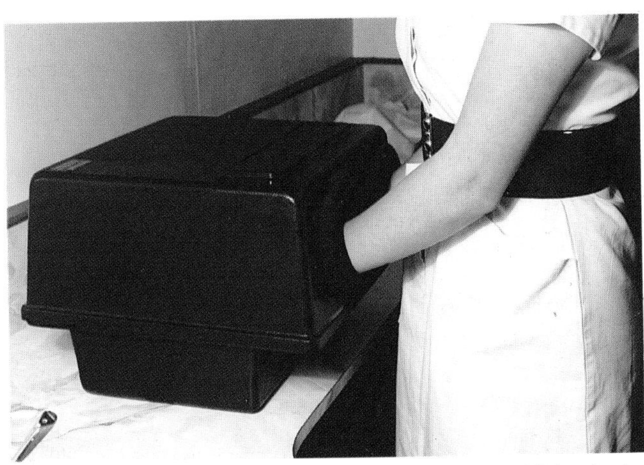

Fig. 6 A manual x-ray processing unit, which uses rapid developing and fixing solutions. It has a viewing panel in the cover and two arm holes.

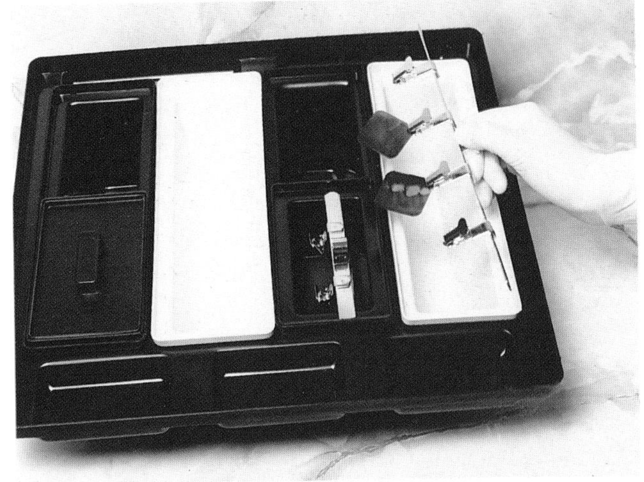

Fig. 7 When the cover is removed, four tanks are revealed. The two small black tanks contain, on the left, developer, with a lid to minimise oxidation, and, on the right, fixing solution. The two white tanks contain water.

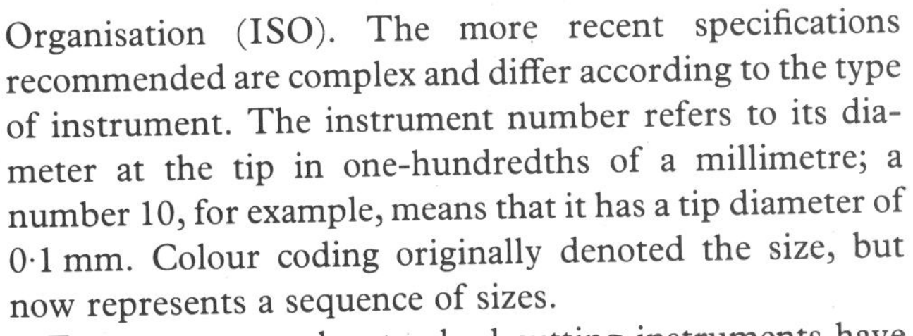

Organisation (ISO). The more recent specifications recommended are complex and differ according to the type of instrument. The instrument number refers to its diameter at the tip in one-hundredths of a millimetre; a number 10, for example, means that it has a tip diameter of 0·1 mm. Colour coding originally denoted the size, but now represents a sequence of sizes.

For many years the standard cutting instruments have been the reamer, K-type file, and Hedstroem file, but several new instruments have recently been introduced (fig. 10).

Reamer

The reamer is constructed from a square or triangular blank which is then twisted into a spiral. The reamer will only cut dentine when it is rotated in the canal; the mode of action that is described for its use is a quarter to a half turn and withdrawal. The stiffness of an instrument increases with each larger size, so that larger reamers in curved canals will tend to cut a wider channel near the apical end of the root canal (apical zipping).

K-type file

These instruments are made from a square or triangular blank, but have more twists to form a tighter spiral than a reamer. The angle of the blades or flutes is consequently nearer a right-angle to the shank than a reamer, so that either a reaming or a filing action may be used. A filing action is more likely to produce an even taper than a reaming action and will not produce apical zipping. It is for this reason that the author no longer recommends reamers.

Hedstroem file

The Hedstroem file is machined from a round tapered blank. A spiral groove is cut into the shank, producing a sharp blade. Only a filing action should be used with this instrument because of the angle of the blade, and there is a possibility of fracture with a reaming action if the blades are engaged in the dentine. The Hedstroem file is useful for removing gutta-percha root fillings (*see* Chapter 11).

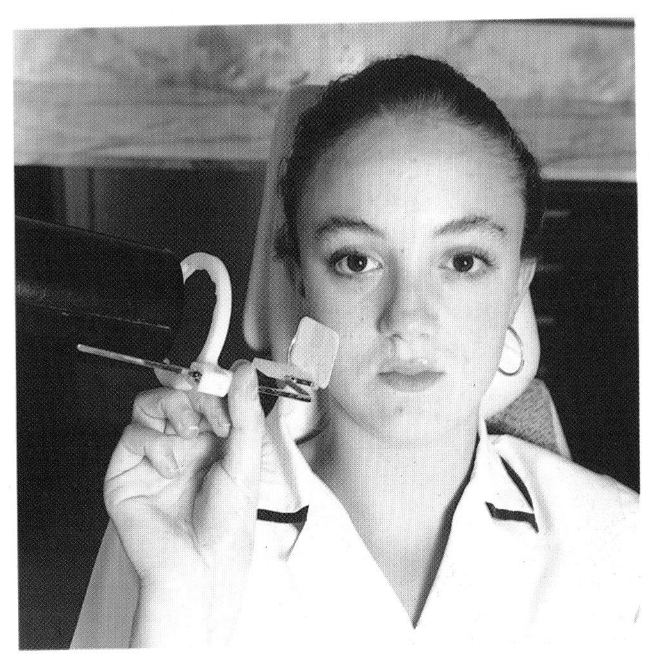

Fig. 8 A film holder for taking parallel radiographs.

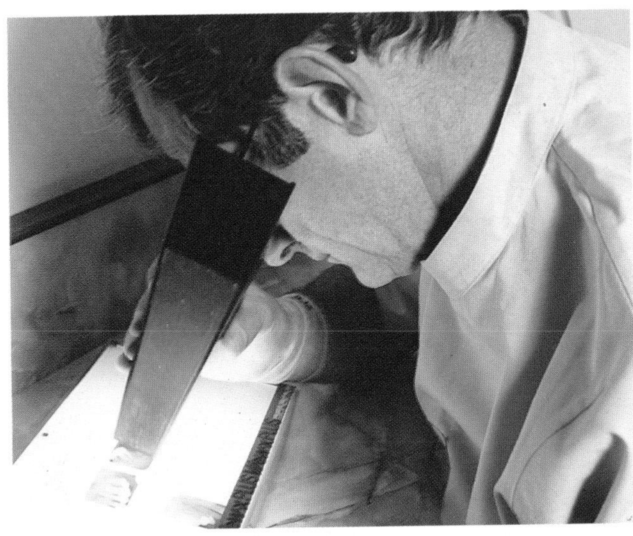

Fig. 9 A portable triangular viewer in use. This magnifies the radiograph and cuts out glare.

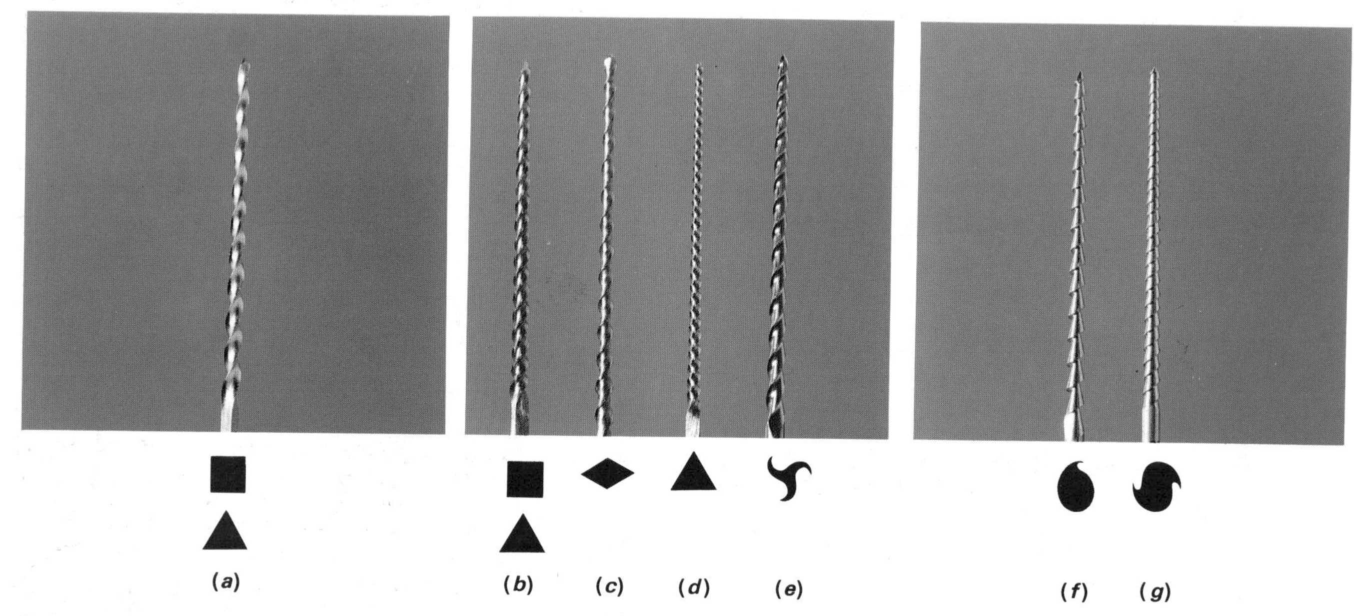

Fig. 10 Hand instruments. (*a*) Reamer; (*b*) K-file; (*c*) K Flex file; (*d*) Flex-o-file; (*e*) Helifile; (*f*) Hedstroem file; (*g*) Unifile. Below is shown a cross-section of each instrument.

Unifile

One of the new breed of instruments, the Unifile appears to be identical to the Hedstroem file. In cross-section, however, the Unifile has two blades, whereas the Hedstroem file has only one. The grooves cut into the shank of the Unifile remain at the same depth throughout. The effect is that the instrument is stiff and less prone to fracture in the coronal and middle thirds of the shank, and is flexible in the apical third, which corresponds to the position of the curve in the majority of roots.

Helifile

The method of manufacture of the Helifile is similar to the Hedstroem and Unifile, except that in cross-section there are three blades. The appearance of the instrument resembles a reamer rather than a Hedstroem file. Little information is available yet concerning its cutting ability or resistance to fracture.

K-flex files

K-flex files are similar to K-files, the difference being the diamond-shaped cross-section. The acute angle of the diamond shape provides the instrument with two sharp blades and the narrower diameter allows greater flexibility in the shaft than a conventional K-file. The manufacturers claim that more debris is collected between the blades and therefore removed from the canal than with a standard K-file.

Flex-o-file

This instrument is manufactured in the same way as a K-file, but using a more flexible type of steel. It does not fracture easily and is so flexible that it is possible to tie a knot in the shank of the smaller sizes.

Power-assisted instruments

A handpiece which provides a mechanical movement to the root canal cutting instrument has been available since 1964. Until recently, the movements have been a reciprocating action through 90° and/or a vertical movement, according to the design and make.

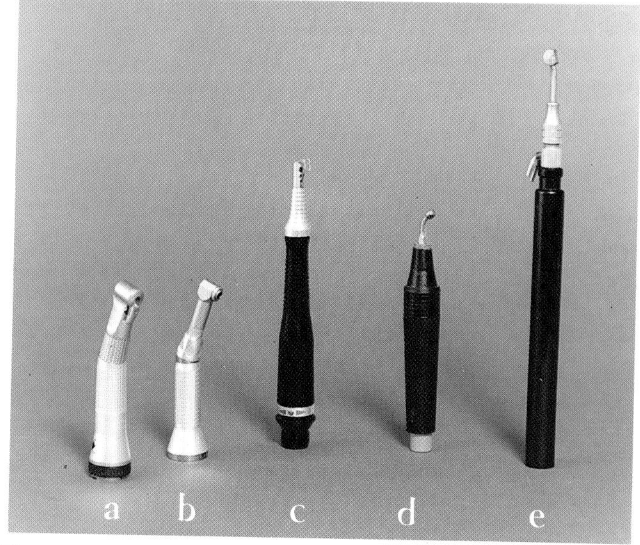

Fig. 11 Power assisted handpieces. (*a*) Giromatic which contra-rotates instruments through 90°. (*b*) Canal finder vibrates instrument but allows free rotation movement. (*c*) Sonic Air, the subsonic vibrations are set to 0·5 mm either side of rest position which is approximately 1500 Hz. (*d*) Piezon Master, an ultrasonic piezoelectric handpiece with a frequency of 28 000 Hz. (*e*) Cavi Endo ultrasonic magnetostrictive handpiece with a frequency of 25 000 Hz.

Several new handpieces are now available, some of which produce a vibratory movement and others ultrasound (fig. 11). There is considerable interest in power-assisted instruments to facilitate cleaning and shaping root canals. Unfortunately, there is a lack of published research to substantiate many of the claims made for these instruments.

For many years the range of root canal cutting instruments has consisted of engine-mounted barbed broaches (giromatic cleansers), Hedstroem files and engine reamers.

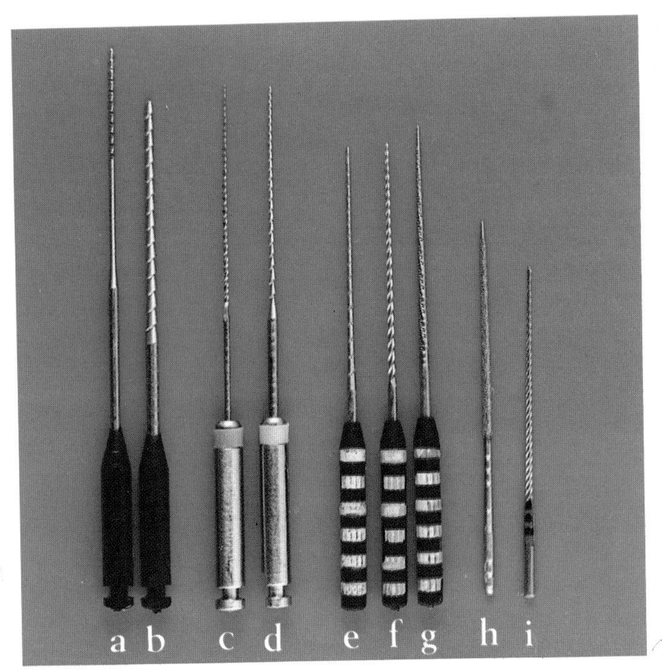

Fig. 12 Power assisted cutting instruments. (*a*) Girocleanser similar to a barbed broach; (*b*) Girofile—A mounted Hedstroem file used in the Giromatic handpiece; (*c*) (*d*) K-file and Hedstroem file for the canal finder system. The Sonic Air uses (*e*) the shaper, (*f*) the helisonic and (*g*) the Rispisonic. (*h*) A diamond file for the Caviendo Ultrasonic System and, finally, (*i*) an ultrasonic K-file.

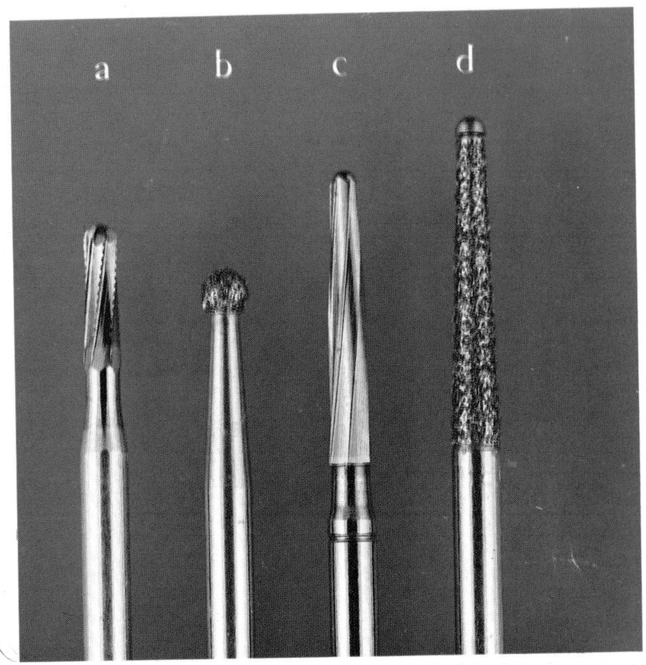

Fig. 13 Access cavity burs. (*a*) Tungsten carbide dome fissure crosscut FG; (*b*) Round diamond FG; (*c*) Tungsten carbide tapered FG non-end cutting; (*d*) Tapered diamond FG non-end cutting.

There are currently four new designs: the Rispi, which is similar to a barbed broach, the Heligirofile and Helisonic, which are for the Giromatic and Sonic Air, the shaper designed by Professor Laurichesse, which resembles a Rispi with smaller barbs, and, finally, a diamond coated file available for the ultrasound handpiece (fig. 12).

Spiral root canal fillers are seldom used in modern endodontics. Their main use is for the insertion of calcium hydroxide into the root canal. When a spiral filler is required, the blade type is preferred by the author, as this is the least likely to fracture. The size selected should fit loosely in the canal.

Burs

Several types of bur may be required for root canal treatment. Some of these are described below.

Cutting an access cavity (fig. 13)

It is generally accepted that high speed burs should be used to gain access and shape the cavity. A tungsten carbide tapered fissure bur is used for initial penetration, except for porcelain crowns when a round diamond bur is used. A tapered safe-ended diamond or tungsten carbide bur is then used so that the roof of the pulp chamber can be removed without damaging the floor.

Location of canal

Burs should only be used as a last resort to locate a sclerosed canal because of the danger of perforation. Small round burs are used; the standard length is usually too short but longer shank burs are available (fig. 14).

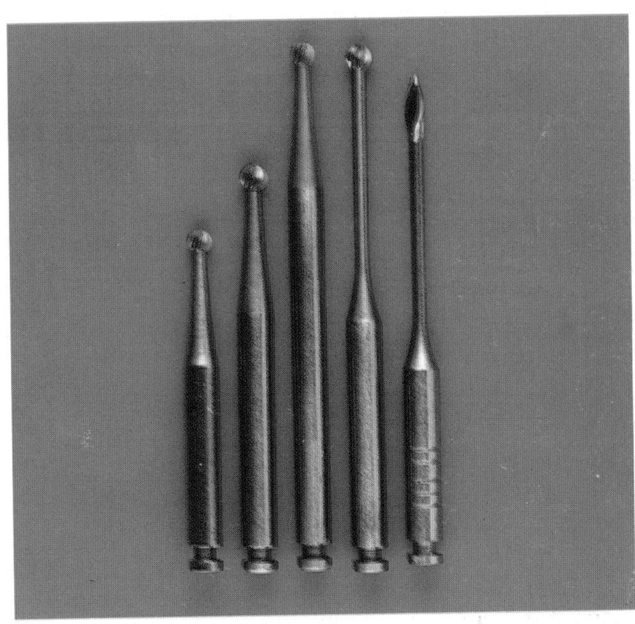

Fig. 14 Canal location. Three round burs with different shank lengths are shown. The goose necked bur, which has a slender shank, a Gates-Glidden bur, which comes in six sizes, is non-end cutting and is used to widen the straight portion of canals.

Canal preparation

The use of rotary cutting instruments in a standard handpiece is condemned because of the danger of fracture of the instrument or perforation of the root canal. The exception to this rule is the Gates-Glidden bur, which has a safe-ended tip. In addition, the site of fracture, if it does occur, is almost always near the hub so the fractured piece is easily removed. The bur is recommended for final

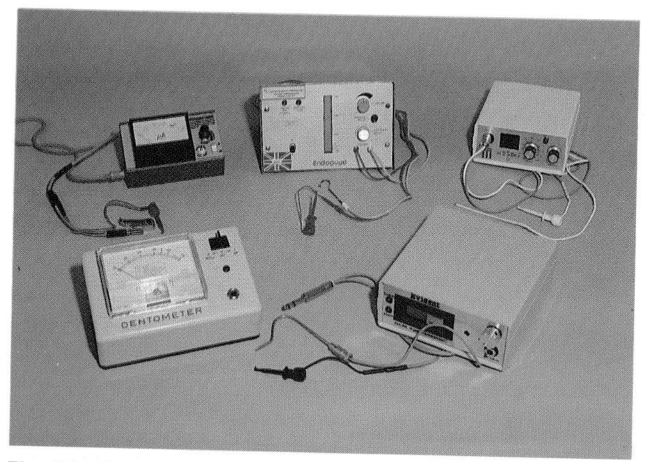

Fig. 15 Electronic apex locators. (*Top row*) Formatron, Endoguyd and Neosono D. (*Bottom row*) Dentometer, Evident RCM.

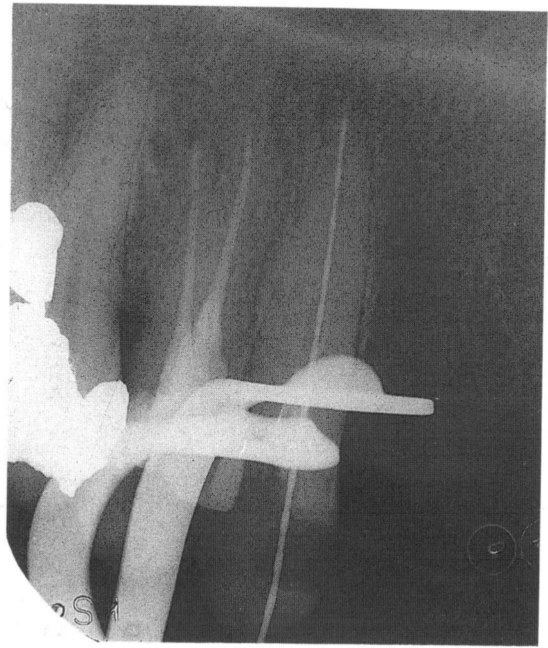

Fig. 16 An instrument used in a diagnostic radiograph to help determine the length of the root canal. The metal part of the shank is a series of gradations, each one 1·0 mm in length. The shank has been covered in a fine layer of silicone so that it has a smooth finish.

flaring of the coronal portion of the canal so that preparation time is reduced. The bur may also be used to make post space and to remove gutta-percha from the canal. Gates-Glidden burs are manufactured in six sizes; the one selected should fit loosely in the canal (fig. 14).

Measurement of working length

The first step in root canal preparation is to measure the working length, and the technique used should be accurate to 0·5 mm. There are two established methods of assessing the length: one by radiography and the other with the use of an electronic device (fig. 15). Both methods will be described in chapter 5.

Another method of calculating the working length from a radiograph requires a special fine diagnostic measuring instrument. The instrument is placed in the canal approximately 1·0 mm short of the working length which is assessed from the pre-operative radiograph. The instrument shows gradations in millimetres which are visible on the radiograph. This is a useful device as any distortion of the radiograph will condense or elongate the gradations (fig. 16).

The second step in measuring is to record and transfer the working length by placing a mark on the shank of each instrument. There are many different gadgets available (fig. 17) for transfer of the working length; the author prefers a metal ruler. There are also different stops for the instrument, the most popular being rubber or silicone stops (fig. 18). One stop is pear shaped, so that in curved canals the point of the pear may be directed towards the curve placed in the instrument. Metal stops may also be used, but there are four different instrument shank sizes so that four varieties of metal stop are required. Marking paste made from a mixture of zinc oxide and petroleum jelly can be used to provide a stop (fig. 18). A complete system, the test handle, is available, whereby the handle of

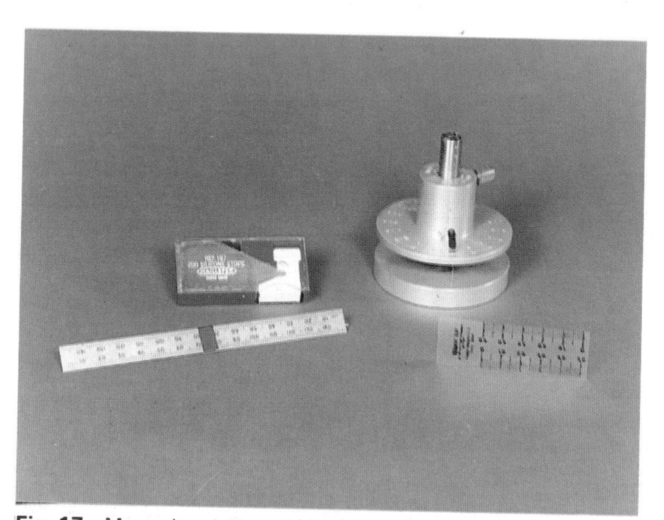

Fig. 17 Measuring devices. *Left to right:* A metal ruler. Box containing silicone stops which has a ruler on the side. An endo gauge and a disposable cellophane ruler.

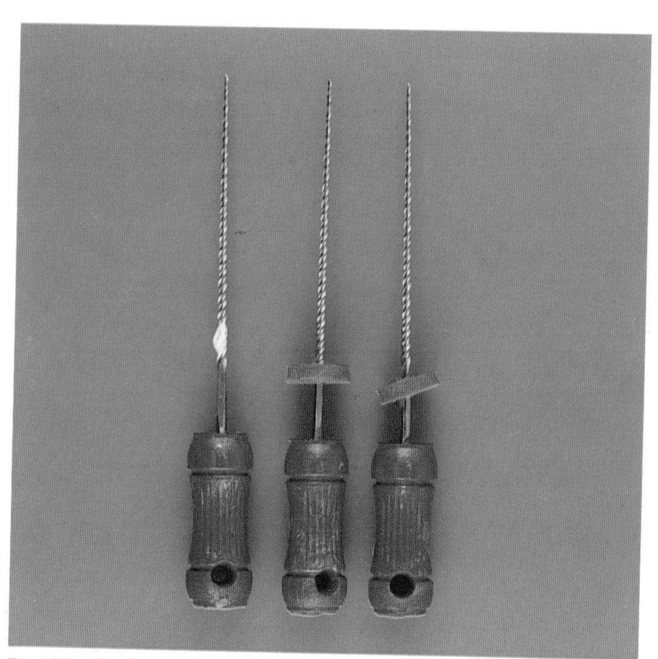

Fig. 18 Marking paste and silicone stops. Care must be taken not to smudge the paste and the silicone stops must be pierced at right angles to give accurate readings.

the instrument is separate and may be tightened at any point along the shank.

Sterilisation

Any instrument which is placed in the mouth should be sterile. Bacteria, viruses, and fungi may contaminate instruments and should be destroyed by chemicals and/or heat.

After use, instruments must be cleaned as soon as possible to remove debris which harbours and protects microorganisms. Cleaning is carried out by scrubbing in warm water and detergent, although most root canal instruments may be cleaned by stabbing them into a sponge. The best method of cleaning is to place the instruments into an ultrasonic bath. The cavitational effects of ultrasonics will dislodge debris from places which are inaccessible to normal cleaning.

When the instruments are clean they must be sterilised. Methods suitable for disinfection and sterilisation in the dental surgery are described below. Boiling water is no longer used for endodontic instruments as it does not kill spores and is unpredictable against viruses.

Chemical

Numerous chemicals are available for disinfection. The one currently recommended is glutaraldehyde, because of its ability to kill spores, and there is indirect evidence that it is effective against the hepatitis virus. Cases of sensitivity to glutaraldehyde have been reported, in which case benzalkonium is an alternative, although it is not effective against spore-forming bacteria or viruses. Chemical disinfection is recommended for items which are not placed in the mouth or which would be damaged by heat, for example burs, rubber dam frames, punches and x-ray film holders (fig. 19).

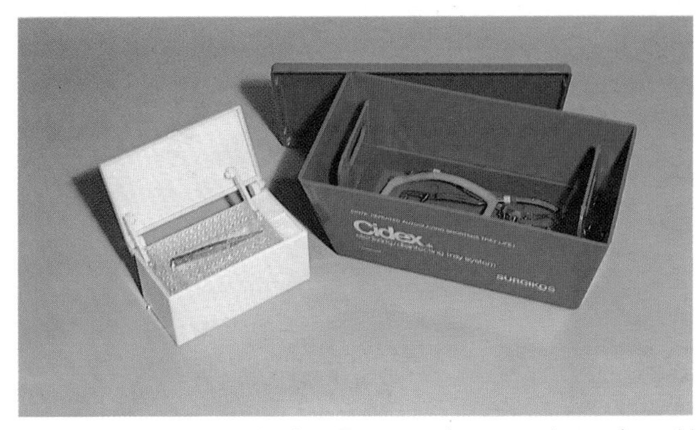

Fig. 19 Chemical sterilisation. Two containers are shown for cold sterilising fluids. The one on the left is for hand instruments. As the lid is closed, the rack is lowered into the fluid. The Cidex bath (*right*) has a removable, perforated tray and is used for larger items.

Fig. 20 Bead sterilisers. The bead steriliser on the left has a temperature gauge, while the Hotty on the right has an indicator light to show when the working temperature has been reached.

Dry heat

Dry heat sterilisers are relatively inexpensive and efficient. The temperature should remain at 160°C for 45 minutes, giving a total cycle time of 90 minutes. Most endodontic instruments may be sterilised with dry heat, although it is not suitable for fabrics, rubber products, paper points and oils.

Moist heat

Microorganisms are destroyed at lower temperatures and in a shorter period in moist heat as all biological reactions are catalysed in water. The autoclave, which is similar to a pressure cooker, operates at a pressure of 15 lb per square inch, so that water boils at 121°C and, provided all air has been displaced, sterilisation occurs in 15 minutes.

The disadvantages of autoclaving are that metal instruments tend to corrode and sharp instruments are dulled. Amine compounds (such as cyclohexylamine) are recommended to reduce corrosion.

Salt or glass bead steriliser

Small instruments such as files or reamers are easily sterilised by this method in 10 seconds. The steriliser should have a temperature gauge to check that the working temperature of 218°C (425°F) has been reached. Common salt will sterilise in a shorter time than glass beads and is recommended (fig. 20).

Storage

The organisation and storage of endodontic instruments must be tailor-made to the operator's individual requirements. Three different systems are suggested.

Test tubes

Standard 11 mm wide Pyrex test tubes accept six reamers or files. The test tubes may be stored in racks (fig. 21) and colour coded aluminium caps placed over the ends. There are four different colours available; each may denote a different length or group of sizes. During root treatment the surgery assistant places the relevant instruments on a sterile sponge (fig. 4). After use, the instruments are cleaned and resterilised inside the test tubes.

Sorter box (fig. 22)

A drawer or shallow box containing a sponge is divided into compartments. The instruments are divided into groups according to size and length. They are placed into the sponge after being cleaned and sterilised. When required they are transferred to a salt bead steriliser and then to a sterile sponge on the instrument tray.

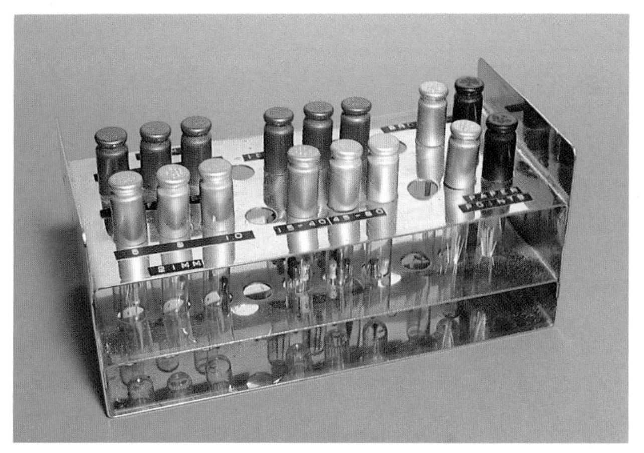

Fig. 21 Test tubes. A simple test tube rack to hold Pyrex test tubes containing groups of instruments. The coloured caps will keep the instruments sterile for a period.

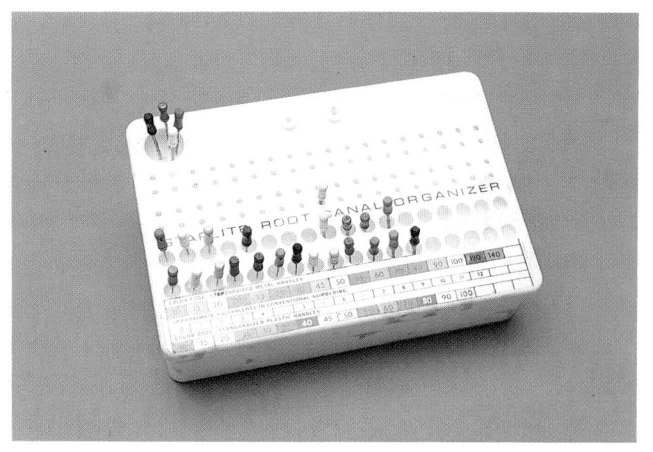

Fig. 22 Sorter box. A manufactured sorter box is shown, but any shallow box or drawer containing a sponge is sufficient.

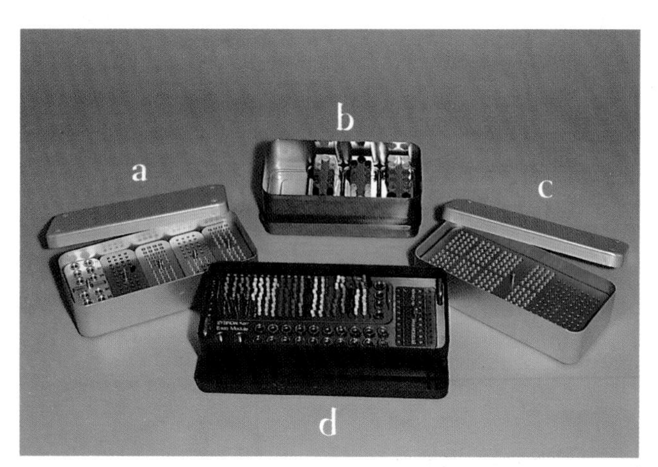

Fig. 23 Sterile boxes. (*a*) Endo-Aide; (*b*) Polybox, which contains up to five endomagazines; (*c*) Boite d'endodontie; (*d*) Endo Module.

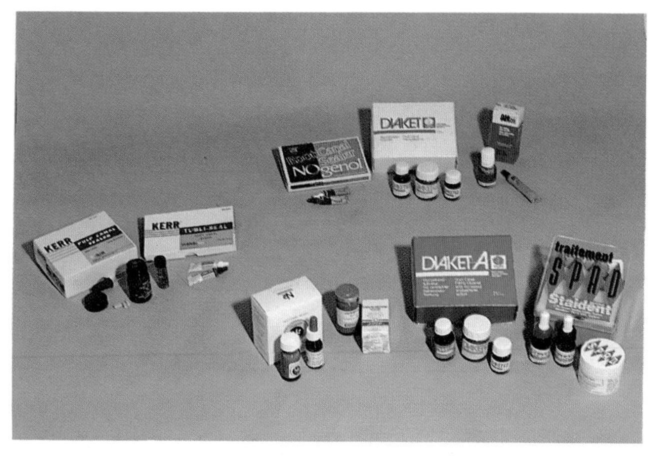

Fig. 24 Sealers. (*Left*) Eugenol: Kerr pulp canal sealer; Tubliseal. (*Top*) Non-eugenol: Nogenol; Diaket; AH26. (*Bottom*) Medicated: N2; Endomethasone; Diaket A; Spad.

Sterile box

Many boxes containing root canal instruments are manufactured so that they can be sterilised (fig. 23). The boxes are metal and have lids so that the contents are kept sterile. Two similar boxes are needed: one of them is filled with the instruments required for one day and sterilised in a hot-air oven; the second box is empty. The instruments are taken from the sterile box and, after use, put into the second box. At the end of the day, the empty spaces are replenished as necessary and the box sterilised ready for the following day.

Sealers

Root canal sealers (fig. 24) may be divided into three groups, according to their main constituents: eugenol, non-eugenol and medicated.

Eugenol

The eugenol-containing group may be divided into sealers based on the Rickert's formula (1931) and those based on Grossman's (1958) (Table I). The essential difference between the two groups is that Rickert's contains precipitated silver and Grossman's has a barium or bismuth salt as the radiopacifier. The disadvantage of Rickert's sealer is that the silver will stain dentine a dark grey. One of the most widely used sealers in this group is Tubliseal, a two-paste system and, consequently, simple to mix; it does not contain silver. Setting time is approximately 20 minutes on the mixing pad and 5 minutes in the root canal.

Non-eugenol sealers

These consist of a wide variety of chemicals. AH26 is an epoxy resin base with a bisphenoldiglycidyl ether liquid. The setting time is 36–48 hours at body temperature. Initially, there is a severe inflammatory response, but this subsides after some weeks. Diaket is a polyketone and is presented as a fine powder and thick viscous liquid. The setting time is 8 minutes on the mixing pad and somewhat quicker in the root canal. The third bottle shown in figure 24 is a solvent used to remove excess sealer from the pulp chamber. A more recent addition to this group is Nogenol, which was developed as a Grossman-type sealer but substituting eugenol, which is a known irritant, for a blander substance. The setting time of Nogenol is approximately 10 minutes at body temperature. Chloropercha, a combination of chloroform and gutta-percha, has fallen into disrepute because of the shrinkage that occurs.

Medicated

The current thinking is that provided the principles of root canal preparation and filling are observed, there is no

Table I Composition of root canal sealers

	Rickert's		Grossman's		Tubliseal		Pulp canal sealer	
Eugenol	*Powder* (g)		*Powder*		*Base*		*Powder*	
	Zinc oxide	41·2	Zinc oxide	42 parts	Zinc oxide		Zinc oxide	
	Precipitated silver	30·0	Staybelite resin	27 parts	Oleo resins		Precipitated molecular silver	
	White resin	16·0	Bismuth subcarbonate	15 parts	Bismuth trioxide		Oleo resins (white resin)	
	Thymol iodide	12·8	Barium sulphate	15 parts	Thymol iodide		Thymol iodide	
			Sodium borate anhydrous	1 part				
	Liquid (ml)		*Liquid*		*Oils and waxes*		*Liquid*	
	Oil of cloves	78·0	Eugenol		Catalyst		Oil of cloves	
	Canada Balsam	22·0			Eugenol		Canada Balsam	
					Polymerised resin			
					Annidalin			
	AH26		**Diaket**		**Nogenol**			
Non-eugenol	Epoxy resin base with bisphenoldiglycidyl ether liquid		*Powder*		Zinc oxide			
			Bismuth phosphate		Barium sulphate			
			Zinc oxide		Natural resin			
			Liquid		Salicylic acid			
			(same as Diaket A except for hexa- and dichlorophene)		Vegetable oil			
					Fatty acids			
	Diaket A (g)		**Endomethasone (g)**		**N2 Universal (%)**		**Sealapex (%)**	
Medicated	*Powder*		Dexamethasone	0·01	*Powder*		Calcium oxide	15·0
	Bismuth phosphate	0·3	Hydrocortisone acetate	1·60	Zinc oxide	*ad* 100	Barium sulphate	20·4
	Zinc oxide	1·0	Thymol iodide	25·00	Paraformaldehyde	4·7	Zinc oxide	6·5
	Liquid		Paraformaldehyde	2·20	Bismuth subcarbonate	6·2	Sub-Micron silica	3·0
	Hexachlorophene	0·05	Radiopaque excipient *q.s.*	100	Bismuth subnitrate	7·3	Titanium dioxide	2·2
	Dichlorophene	0·005			Titanium dioxide	18·4	Zinc stearate	1·0
	Triethanolamine	0·002			*Liquid*		In a blend of ethyl toluene sulphonamide, poly (methylene methyl salicylate) resin, Isobutyl salicylate and a pigment.	
	Propionylacetophenone	0·72			Eugenol	*ad* 100		
	Co-polymers of vinyl acetate, vinyl chloride, vinyl isobutyl, ether	1·0			Oleum rosae	1·8		
					Oleum lavandulae	1·7		

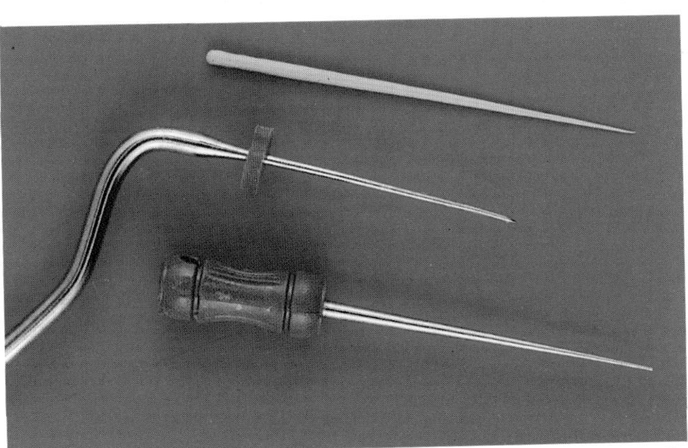

Fig. 25 Lateral condensation. A finger spreader and the head of a long-handled spreader are shown. The accessory gutta-percha cone is approximately the same size as the spreaders.

justification for the use of therapeutic sealers. The active ingredient in the majority of medicated sealers is paraformaldehyde, which is usually accompanied by a corticosteroid.

Diaket A is identical to Diaket, except that it contains hexachlorophene and dichlorophene as active ingredients, which makes this sealer potentially the least offensive to the periapical tissues. N2, which used to consist of two different sealers, apical and normal, is now marketed as N2 Universal, and the formula has been altered by the removal of hydrocortisone, prednisolone and barium sulphate. N2 has, to a large extent, been replaced by Endomethasone, although the composition of both sealers is similar. The active ingredient of Endomethasone is paraformaldehyde, although on the packaging this is referred to as trioxymethylene.

Calcium hydroxide sealers have been introduced on to the market and early reports are promising. The author has a slight reservation that if the calcium and hydroxyl ions are freely available, it means the sealer is soluble, yet if they are chemically bound in the sealer there would be little benefit. Sealapex is a most useful addition to the range of sealers.

Root canal filling instruments

Lateral condensation using gutta-percha requires either long-handled or finger spreaders (fig. 25).

Spreader

This has a long, tapered shank with a sharp point. The instrument is used to force gutta-percha laterally against the walls of the root canal and provide a space for the insertion of further gutta-percha points. There are several sizes available, and these are selected according to the canal size and the size of the gutta-percha point. Both long-handled spreaders and finger spreaders are available; the choice depends on personal preference. The advantage of finger spreaders is that less force can be used, and this reduces the risk of root fracture.

Reference

1 Reuter J. The isolation of teeth and the protection of the patient during endodontic treatment. *Int Endod J* 1983; **16:** 173–181.

Hand instruments

K-flex: Kerr Manufacturing Company, Detroit 8, Michigan, USA
Flex-o-file: Les Fils d'Auguste, Maillefer SA, Ballaigues, CH-1338, Switzerland
Unifile: Ransom and Randolf Dentsply, 324 Chestnut Street, Toledo, Ohio 43604, USA
Helifile: Micromega SA, 5–2 rue de Tunnel, Besançon 25006, France

Power-assisted instruments

Giromatic: Micromega SA, 5–2, rue de Tunnel, Besançon 25006, France
Sonic Air: Micromega SA, 5–2, rue de Tunnel, Besançon 25006, France
Cavi Endo: De Trey Dentsply Ltd, Weybridge, Surrey
Piezon Master: E.M.S. Optident, Castlefields Lane, Bingley, West Yorks
Hawes-Neos root fillers: 6925 Gentillino, Lugano, Switzerland
Canal finder system: Société Endotechnic, 38 ch St Jean du Désert, 13005 Marseilles, France
Girocleanser, Girofile, Shaper, Rispisonic and Helisonic: Micromega SA, 5–2 rue de Tunnel, Besançon 25006, France

Burs

IFG 1958 Jet carbide: Beaver Dental Products, Morrisburg, Ontario, Canada
FG 332/018 Horico: Hopf, Ringleb and Company, GmbH & Cie, Gardeschutzenweg 82, D-1000 Berlin 45, W. Germany
Non end cutting TC: 152 FG Endo-2, Maillefer SA, Baillagues, Switzerland
Goose neck or Muller bur: Hager and Meisinger GmbH, Kronprinzenstrasse 5–11, D-4000 Dusseldorf 1, W. Germany
Gates-Glidden: Hager and Meisinger GmbH, Kronprinzenstrasse 5–11, D-4000 Dusseldorf 1, W. Germany

Measurement of working length

Endoguyd: Scalatron, 57 Lee High Road, Lewisham, London SE13 5NS
Neosono D: Amadent, PO Box 3422, Cherry Hill, NJ 08034, USA
Dentometer: Dahlin Electromedium, Copenhagen, Denmark
RCM: Evident, 57 Wellington Court, Wellington Road, London NW8 9TD
Formatron: Parkell Electronics Division, Farmingdale, NY, USA
Endogauge: KRD, Manchester M3 2AD
Endostops silicone: Polydent SA, CH-6607, Taverne, Switzerland
Vari-fix metal stops and gauge: Vereinigte Dentalwerke, Anteos Beutelroc, Zipperer, Zardsky Ehrler KG, D8 Munich 70, W. Germany
Test handle system: Vereinigte, Dentalwerke, Anteos Beutelroc, Zipperer, Zardsky Ehrler KG, D8 Munich 70, W. Germany
Red rubber stops: Micromega SA, 5–2 rue de Tunnel, Besançon 25006, France
Endometer: Toringe Consult AB, Satsjöbaden, Sweden
Diagnostic measuring device: Maillefer SA, Baillagues, Switzerland

Radiography

Manual processing unit: Manumatic, Westone Products Ltd, 104–112 Marylebone Lane, London W1M 5FU
Rapid developing chemicals: RXD-30 and RXF-30, Westone Products Ltd, 104–112 Marylebone Lane, London W1M 5FU
X-ray film holder, Rinn holder: Nesor Products Ltd, Claremont Mall, Pentonville Road, London N1 9HR

Sterilisation

Sonicleaner: Dawe Instruments, Concord Road, Western Avenue, London W3 0SD
Sporocidin (glutaraldehyde): The Sporocidin Company, 4000 Massachusetts Avenue NW, Washington DC, USA
Cidex (glutaraldehyde): Johnson & Johnson, 260 Bath Road, Slough, Buckinghamshire SL1 4EA
Roccal (benzalkonium): Winthrop Laboratories, Surbiton upon Thames, Surrey
Bead steriliser: National Keystone Products, 2nd and Noble Street, Philadelphia, PA 19123, USA
Hotty salt bead steriliser: Produkta, 20145 Milano, via Canova 33, Italy
Sterilisation pouch: Code P5, DRG Supplies, Dixon Road, Brislington, Bristol BS4 5QY
Portable triangular viewer: Dental Rontgen, Malmofilialen, Box 30110, 200–16, Malmo, Sweden

Storage

Test tubes: Solmedia, 31 Orford Road, London E17 9NL
Cap-o-test tops: Solmedia, 31 Orford Road, London E17 9NL
Sorter box: Star Dental, Philadelphia, PA 19428, USA
Kerr endo-module: Kerr Division of Sybron (Europe) AG, Aeschengrabenio, CH-4010, Basle, Switzerland
Endo Aide: France Dentaire, 21 Bis Rue, Faubourg, Montmartre 750009, Paris, France
Boîte d'endodontie: France Dentaire, 31 Bis Rue, Faubourg, Montmartre 750009, Paris, France
Poly box: Vereinigte Dentalwerke, Anteos Beutelroc Zipperer, Zardsky Ehrler KG, D8 Munich 70, W. Germany

Intracanal medication

Sodium hypochlorite household bleach: Sainsbury's supermarket (England); Chlorox in USA
Ledermix: Lederle Laboratories, Cyanamid of GB Ltd, Fareham Road, Gosport, Hampshire PO13 0AS

Basic instruments and materials

Instrument pack

Front Surface Mirror: India head: Union Broach Corporation, 36–40, 37th Street, Long Island City, NY, 11101, USA
Endolocking tweezers: Carl Martin, PO Box 170148, D 565 Solingen 17, Hohscheid, W. Germany
DG 16: Hu Friedy, 3232 North Rockwell Street, Chicago, Illinois 60618, USA
Excavator Ash 141/142: Amalgamated Dental Co Ltd, 26–40 Broadwick Street, London W1A 2AD
Marquis Perio Probe: Prima, 23 Faris Barn Drive, Woodham, Weybridge, Surrey
American pattern probe No 3: Prima, 23 Faris Barn Drive, Woodham, Weybridge, Surrey
Briault probe: Ash Instruments, Dentsply Ltd, Weybridge, Surrey
Paper points: Dental Supply and Engineering Co Ltd, High Path, London SW19 2LA
Patient bib: Kent Dental Ltd, Wested Lane, Swanley, Kent BR8 8TE

Trays

RAF Tray: J & S Davis, Cordent House, 34/36 Friern Park, London N12 9DQ
Double tray (DT): J & S Davis, Cordent House, 34/36 Friern Park, London N12 9DQ

Rubber dam

Black rubber dam: The Hygienic Corporation, 1245 Home Avenue, Akron, Ohio 44310, USA
Green rubber dam: Associated Dental Products Ltd, Kemdent Works, Swindon, Wiltshire SN5 9HT
Starlite visi-frame: Star Dental Mfg, Philadelphia, PA 19428, USA
Forceps: Ivory, 660 Jessie Street, San Fernando 91340 CA, USA
Punch: Ivory, 660 Jessie Street, San Fernando 91304 CA, USA
Clamps: Ivory, 660 Jessie Street, San Fernando 91340 CA, USA

Sealers

Diaket & Diaket A: Espe GmbH, Seefeld, Oberbayern, W. Germany

AH 26: De Trey Dentsply Ltd, Weybridge, Surrey
N2 Universal: Indrag AGSA SA, CH-6616, Lausanne, Switzerland
Nogenol: COE Laboratories, Chicago, Illinois 60618, USA
Spad: BP7, 21801 Quetigny Lef, Dijon, Cedex, France
Endomethasone: Specialités Septodont, 58 rue du Pont de Creteil, 94100 Saint-Maur, France
Kerr pulp canal sealer: Kerr Manufacturing Company, Detroit 8, Michigan, USA
Tubliseal: Kerr Manufacturing Company, Detroit 8, Michigan, USA
Sealapex: Kerr Manufacturing Company, Detroit 8, Michigan, USA

Root canal filling instruments

5/7 Plugger: Hu Friedy, 3232 North Rockwell Street, Chicago, Illinois 60618, USA
Finger spreader and plugger: Kerr Manufacturing Company, Detroit 8, Michigan, USA
No. 3 spreader: Kerr Manufacturing Company, Detroit 8, Michigan, USA
No. 8 plugger: Kerr Manufacturing Company, Detroit 8, Michigan, USA
PCA D4: Pulpdent Corporation of America, 75 Boylston Street, Brooklyn, Massachusetts 02147, USA
D11T: Star Dental, Philadelphia, PA 19428, USA

4

Treatment of Endodontic Emergencies

The aim of emergency endodontic treatment is to relieve pain and control any inflammation or infection that may be present. Although insufficient time may prevent ideal treatment from being carried out, the procedures followed should not prejudice any final treatment plan. It has been reported that nearly 90% of patients seeking emergency dental treatment have symptoms of pulpal or periapical disease.[1,2]

Patients who present as endodontic emergencies can be divided into three main groups.

Before treatment:

(1) Pulpal pain;
 (a) Reversible pulpitis;
 (b) Irreversible pulpitis;
(2) Acute periapical abscess;
(3) Cracked tooth syndrome.

Patients under treatment:

(1) Recent restorative treatment;
(2) Periodontal treatment;
(3) Exposure of the pulp;
(4) Fracture of the root or crown;
(5) Pain as a result of instrumentation;
 (a) Acute apical periodontitis;
 (b) Phoenix abscess.

Post-endodontic treatment:

(1) High restoration;
(2) Overfilling;
(3) Root filling;
(4) Root fracture.

Before treatment

Details of the patient's complaint should be considered together with the medical history. The following points are particularly relevant and are covered more fully in Chapter 2.

(1) Where is the pain?
(2) When was the pain first noticed?
(3) Description of the pain.
(4) Under what circumstances does the pain occur?
(5) Does anything relieve it?
(6) Any associated tenderness or swelling.
(7) Previous dental history:
 (a) recent treatment, (b) periodontal treatment, (c) any history of trauma to the teeth.

Particular note should be made of any disorders which may affect the differential diagnosis of dental pain, such as myofascial pain dysfunction syndrome (MPD), neurological disorders such as trigeminal neuralgia, vascular pain syndromes and maxillary sinus disorders.

Diagnostic aids

(1) Periapical radiographs taken with a paralleling technique.
(2) Electric pulp tester for testing pulpal responses.
(3) Ice sticks, hot gutta-percha and hot water for testing thermal responses.
(4) Periodontal probe.

Pulpal pain

The histological state of the pulp cannot be assessed clinically.[2–4] Nevertheless, the signs and symptoms associated with progressive pulpal and periapical disease can give a reasonable indication of the likely state of an inflamed pulp, that is whether it is reversibly or irreversibly damaged.[5]

Irritation of the pulp causes inflammation, and the level of response will depend on the severity of the irritant. If it is mild, the inflammatory process may resolve in a similar fashion to that of other connective tissues; a layer of reparative dentine may be formed as protection from further injury. However, if the irritation is more severe, with extensive cellular destruction, further inflammatory changes involving the rest of the pulp will take place, which could eventually lead to total pulp necrosis.

There are features of pulpitis which can make the borderline between reversible and irreversible pulpitis difficult to determine clinically. In general, if the responses to several tests are exaggerated, then an irreversible state is possible.

The essential feature of a reversible pulpitis is that pain ceases as soon as the stimulus is removed, whether it is caused by hot or cold fluids, or sweet food. The teeth are not tender to percussion, except when occlusal trauma is a factor. Initially, the following treatment may be all that is necessary.

(1) Check the occlusion and remove non-working facets.
(2) Place a sedative dressing in a cavity after removal of deep caries.
(3) Apply a fluoride varnish or Tresiolan* to sensitive dentine and prescribe a desensitising toothpaste.

*Details can be found at the end of the chapter.

Should the symptoms persist and the level of pain increase in duration and intensity, then the pulpitis is likely to be irreversible. The patient may be unable to decide which tooth is causing the problem, since the pain is often referred to teeth in both the upper and lower jaw on the same side. In the early stages, the tooth may exhibit a prolonged reaction to both hot and cold fluids, but is not necessarily tender to percussion. Only when the inflammation has spread throughout the pulp and has involved the periodontal ligament, will the tooth become tender to bite on. In these circumstances, the application of heat will cause prolonged pain which may be relieved by cold. Both hot and cold can precipitate a severe bout of pain, but as a rule heat tends to be more significant.

Pain from an irreversibly damaged pulp can be spontaneous and may last from a few seconds to several hours. A characteristic feature of an irreversible pulpitis is when a patient is woken at night by toothache. Even so, a symptomatic pulpitis may become symptomless and pulp tests may give equivocal results. In time, total pulp necrosis may ensue, without the development of further symptoms and the first indication of an irreversibly damaged pulp may be seen as a periapical rarefaction on a radiograph, or the patient may present with an acute periapical abscess.

To summarise, therefore, in reversible pulpitis:

(1) The pain is of very short duration and does not linger after the stimulus has been removed.
(2) The tooth is not tender to percussion.
(3) The pain may be difficult to localise.
(4) The tooth may give an exaggerated response to vitality tests.
(5) The radiographs present with a normal appearance, and there is no apparent widening of the periodontal ligaments.

In irreversible pulpitis:

(1) There is often a history of spontaneous bouts of pain which may last from a few seconds up to several hours.
(2) When hot or cold fluids are applied, the pain elicited will be prolonged. In the later stages, heat will be more significant; cold may well relieve the pain.
(3) Pain may radiate initially, but once the periodontal ligament has become involved, the patient will be able to locate the tooth.
(4) The tooth becomes tender to percussion once inflammation has spread to the periodontal ligament.
(5) A widened periodontal ligament may be seen on the radiographs in the later stages (fig. 1).

Careful evaluation of a patient's dental history and of each test is important. Any one test on its own is an insufficient basis on which to make a diagnosis. Records and radiographs should first be checked for any relevant information such as deep caries, pinned restorations, and the appearance of the periodontal ligament space (fig. 1). Vitality tests can be misleading, as various factors have to be taken into account. For example, the response in an older person may differ from that in someone younger due to secondary dentine deposition and other atrophic changes in the pulp tissue. Electric pulp testing is simply an indication of the presence of vital nerve tissue in the root canal system only and not an indication of the state of health of the pulp tissue.

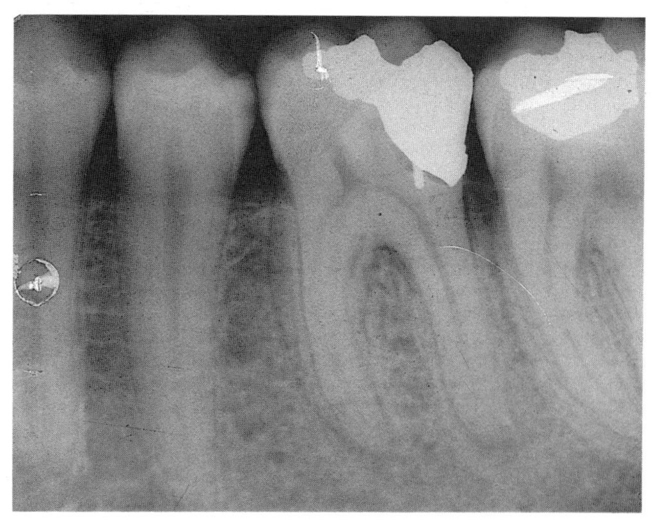

Fig. 1 Initial radiographic assessment. Radiographs should first be checked for any relevant information such as deep caries, pinned restorations, and the appearance of the periodontal ligament space.

Once pulpal inflammation has spread to the periodontal ligament, the resulting inflammatory exudate may cause extrusion of the tooth, which is tender to bite on. This particular symptom, acute apical periodontitis, may be a consequence of occlusal trauma; the occlusion must therefore always be checked.

Ideally, the treatment for irreversible pulpitis is pulp extirpation followed by cleaning and preparation of the root canal system. If time does not permit this, then removal of pulp tissue from the pulp chamber and from the coronal part of the root canal is often effective. Irrigation of the pulp chamber using a 2·5% solution of sodium hypochlorite before carrying out any instrumentation is important (1 part Sainsbury's domestic household bleach to 1 part purified water BP). Sodium hypochlorite (0·5–5·0% solution) has proved to be one of the most effective disinfecting agents used in endodontic treatment.[6] The pulp chamber and root canals are dried, and a dry sterile cotton wool pledget placed in the pulp chamber with a temporary filling to seal the access cavity. Antiseptic solutions such as Cresatin or corticosteroid/antibiotic preparations on cotton wool pledgets have been advocated, but their effectiveness is of doubtful value. Corticosteroid dressings should be used sparingly as there is evidence that suppression of an inflammatory response by steroids allows bacteria to enter the blood stream with ease.[11] This is a particularly undesirable effect in patients who, for example, have a history of rheumatic fever. Studies have shown that provided the pulp chamber and the root canals have been cleansed and dried, medication of the pulp chamber and root canals is of little practical benefit. Paper points are used to dry the canals and under no circumstances should they be left in the canal, otherwise any fluid that enters the canal system will be absorbed and so provide an effective culture medium for any residual bacteria.

Difficulty with local analgesia is a common problem with an acutely inflamed pulp. In addition to standard

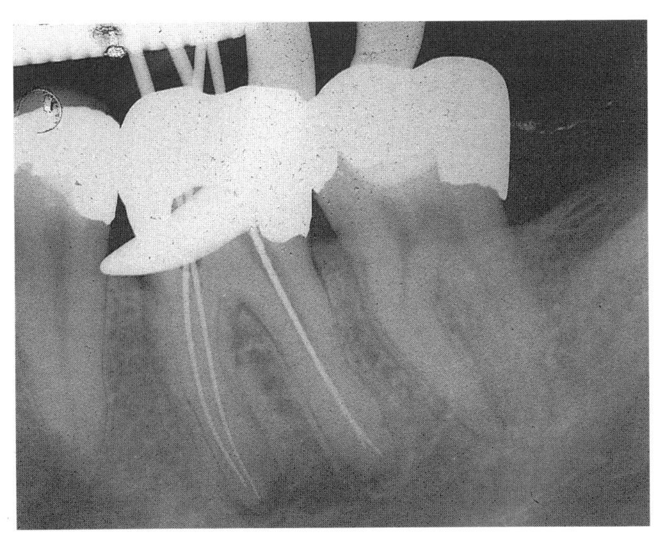

Fig. 2 Acute periapical abscess. Radiographic changes range from a widening of the periodontal ligament space . . .

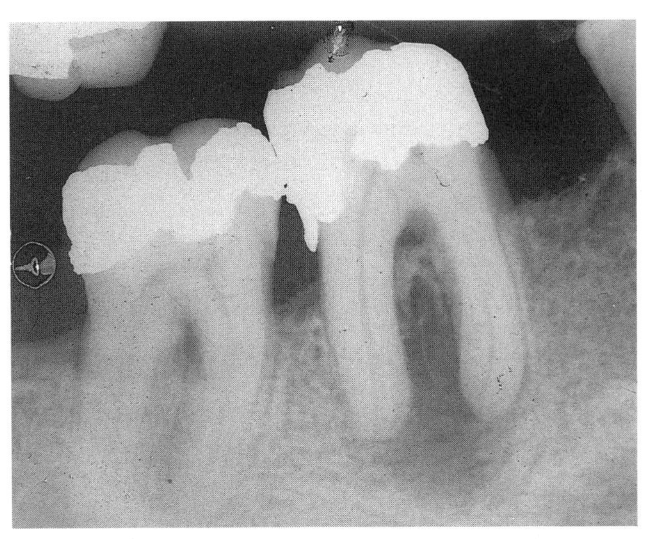

Fig. 3 Acute periapical abscess . . . to a well defined area.

techniques, supplementary analgesia can be obtained with the following:

(1) Intraligamental (intra-osseous) injection.
(2) Intrapulpal analgesia.
(3) Inhalational sedation with local analgesia.

Should these techniques give only moderate success, it is advisable to dress the pulp to allow the inflammation to subside and to postpone pulp extirpation. Cresatin, or a corticosteroid/antibiotic preparation on a cotton wool pledget with a zinc oxide/eugenol dressing will provide an effective, short-term dressing.

Continuation of pain following pulp extirpation may be due to one of the following causes:

(1) The temporary filling is high.
(2) Infected pulp tissue is present in the canal.
(3) Some of the canal contents have been extruded through the apex.
(4) Overinstrumentation of the apex or perforation of the canal wall.
(5) An extra canal may be present which has not been cleaned.

Whatever the cause, the remedy is to irrigate the pulp chamber and root canal system again with 2·5% sodium hypochlorite solution and perhaps gently instrument, then dry and redress the tooth as before.

Acute periapical abscess

This condition develops from an acute periapical periodontitis. In the early stages, the difference between the two is not always clear. Radiographic changes range from a widening of the periodontal ligament space (fig. 2), to a well defined area (fig. 3). The typical symptoms of an acute periapical abscess are a pronounced soft tissue swelling and an exquisitely tender tooth. Extrusion from the socket will often cause the tooth to be mobile. Differential diagnosis of a suspected periapical swelling is important, in case the cause is a lateral periodontal abscess. The diagnosis can be made by testing the vitality of the tooth. If it is vital, then the cause may well be periodontal in origin.

The immediate task is to relieve pressure by establishing drainage, and in the majority of cases this can be achieved by first opening up the pulp chamber. If a soft tissue swelling is present and pointing intra-orally, then it should be incised to establish drainage as well. Initially, gaining access can be difficult because the tooth is often extremely tender. Gently grip the tooth and use a small round diamond bur in a turbine to reduce the trauma of the operation. Regional analgesia may be necessary, and inhalation sedation can prove invaluable.

Blocked canals or the presence of a cellulitis may result in little or no drainage. If a cellulitis is present, medical advice must be sought before any treatment is carried out. However, once access and initial drainage have been achieved, a rubber dam should be applied to complete the operation. Before any further instrumentation is carried out, the pulp chamber should be thoroughly irrigated with a 2·5% solution of sodium hypochlorite to remove as much superficial organic and inorganic debris as possible. This in itself may bring considerable pain relief and will make subsequent instrumentation easier. Having thoroughly debrided the canals with frequent changes of irrigant, the canals should be dried with paper points and a dry sterile cotton wool pledget placed in the pulp chamber. The access cavity is then sealed to prevent re-infection of the canals from the oral cavity. The only exception to this rule is where drainage is profuse and likely to continue for some time; the tooth may then be left open to drain for a maximum period of 24 hours. The patient should be seen the following day for review and a change of dressing if necessary.

Cracked tooth syndrome (posterior teeth)

Crazing of the enamel surface is a common finding on teeth as a consequence of function, but on occasion it may indicate a cracked tooth. If the crack runs deep into dentine and is therefore a fracture, chewing may be painful. Initially, this may not be of sufficient intensity for the patient to seek treatment. However, once the fracture line communicates with the pulp, pulpitis will ensue. A quiescent period of several months may follow before any further symptoms develop. The patient may present with a whole range of bizarre symptoms, many of which are similar to those of irreversible pulpitis:

(1) Pain on chewing.
(2) Sensitivity to hot and cold fluids.
(3) Pain which is difficult to localise.
(4) Pain referred along to the areas served by the fifth cranial nerve.
(5) Acute pulpal pain.
(6) Alveolar abscess.

Diagnosis can be difficult and much depends on the plane of the fracture line and its site on the tooth.[10] Radiographs are unlikely to reveal a fracture unless it runs in a buccolingual plane. A fibre-optic light is a useful aid as it will often reveal the position of the fracture. One diagnostic test is to ask the patient to bite on a piece of folded rubber dam. Care must be exercised as this test may extend the fracture line. The extent of the fracture line and its site will decide whether the tooth can be saved or not. If it is a vertical fracture, involves the root canal system and extends below the level of the alveolar crest, then the prognosis is poor and extraction is indicated. However, if the fracture line is horizontal or diagonal and superficial to the alveolar crest, then the prognosis may be better.

Patients under treatment

Recent restorations

Pain may be a result of:

(1) high filling;
(2) microleakage;
(3) micro-exposure of the pulp;
(4) thermal or mechanical injury during cavity preparation or an inadequate lining under metallic restorations;
(5) chemical irritation from lining or filling materials;
(6) electrical effect of dissimilar metals.

It is not always possible to know beforehand whether there is a pre-existing pulpal condition when operative procedures are undertaken. Consequently, a chronic pulpitis may be converted into an acute pulpitis.

Periodontal treatment

There is always a chance that some of the numerous lateral canals that communicate with the periodontal ligament are exposed when periodontal treatment is carried out. This aspect is considered in the section on 'perio-endo lesions'.

Exposure of the pulp

If a carious exposure is suspected, then removal of deep caries should be carried out under rubber dam. The decision to extirpate the pulp or simply carry out a pulp capping procedure depends on whether the pulp has been irreversibly damaged or not. If there is insufficient time, or any difficulty is experienced with analgesia, temporary treatment, as recommended for irreversible pulpitis, may be carried out.

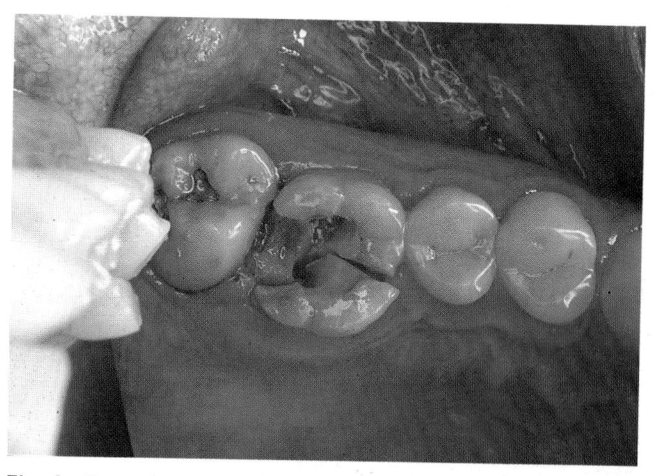

Fig. 4 Root or crown fractures. Most root or crown fractures can be avoided by adequately protecting the tooth during a course of endodontic treatment. If the tooth happens to fracture in a vertical plane, the prognosis is poor and the tooth may have to be extracted.

Root or crown fractures

Most root or crown fractures can be avoided by adequately protecting the tooth during a course of endodontic treatment. If the structure of the tooth is damaged between appointments, pain is likely to occur as a result of salivary and bacterial contamination of the root canal. If the tooth happens to fracture in a vertical plane, the prognosis is poor and the tooth may have to be extracted (fig. 4). In the case of multi-rooted teeth, it may be possible to section the tooth and remove one of the roots.

Pain as a result of instrumentation

The two conditions that may require emergency treatment during a course of endodontic treatment are:

(1) acute apical periodontitis;
(2) Phoenix abscess.

Acute apical periodontitis may arise as a result of over-instrumentation, extrusion of the canal contents through the apex, leaving the tooth in traumatic occlusion, or placing too much medicament in the pulp chamber as an inter-appointment dressing.

Irrigation of the canal with sodium hypochlorite and careful drying with paper points is usually sufficient to alleviate the symptoms. The occlusion must be checked, as there is likely to be a certain amount of extrusion of the tooth from its socket.

Phoenix abscess can be one of the most troublesome conditions to deal with and occurs after initial instrumentation of a tooth with a pre-existing chronic periapical lesion (fig. 5). The reasons for this phenomenon are not fully understood, but it is thought to be due to an alteration of the internal environment of the root canal space during instrumentation which activates the bacterial flora. Research has shown that the bacteriology of necrotic root canals is more complex than was previously thought, in particular the role played by anaerobic organisms.[7–9] Treatment consists of irrigation, debridement of the root canal and establishing drainage. In severe cases, it may be necessary to prescribe an antibiotic.

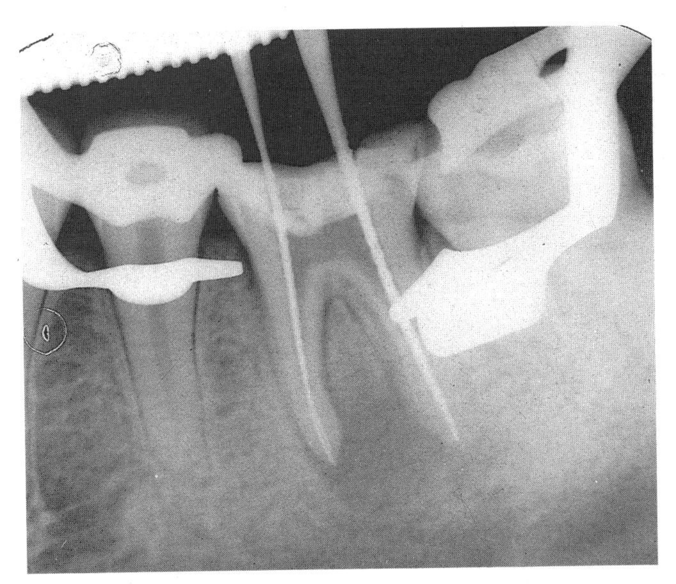

Fig. 5 Phoenix abscess. This can be one of the most troublesome conditions to deal with and occurs after initial instrumentation of a tooth with a pre-existing chronic periapical lesion.

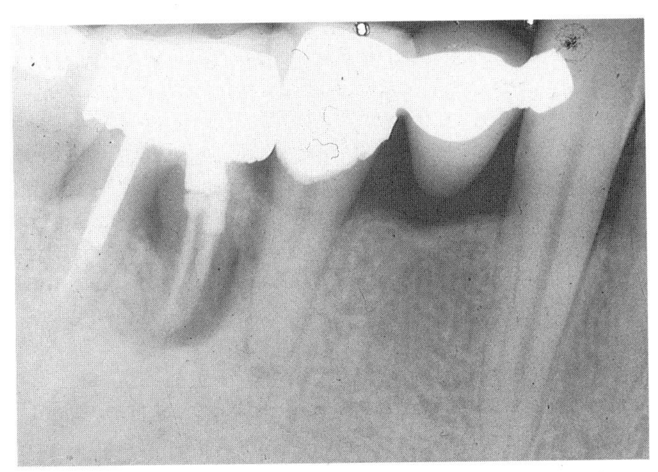

Fig. 6 Fracture of the mesial root of a lower first molar. The tooth was divided buccolingually and the mesial root extracted. The distal root was subsequently found to be fractured and was extracted as well.

Post-endodontic treatment

The following factors need to be considered should pain occur following sealing of the root canal system:

(1) High restoration.
(2) Overfilling.
(3) Underfilling.
(4) Root fracture.

Once obturation of the root canal space has been completed, restoration of the rest of the tooth can be carried out. The occlusion must be checked for interferences, to avoid an apical periodontitis, or worse, a fractured tooth.

Root fillings that are apparently overfilled do not as a rule cause more than mild discomfort after completion. The most likely cause of pain following obturation of the root canal space is the presence of infected material in the periapical region. The significance of an underfilled root canal is whether the canal has been properly cleaned and prepared in the first instance, and infected debris is still present in the canal. Post-endodontic pain in these circumstances may well be due to inadequate debridement of the canal.

Removal of an overextended root filling is rarely completely successful and the options left are as follows:

(1) Prescription of analgesics and, if the pain is more severe and infection is present, antibiotics.
(2) Removal of the root filling and repreparation of the root canal.
(3) Apicectomy.

Root fracture

The forces needed to condense a root filling, using the lateral condensation of gutta-percha technique, should not be excessive; too much pressure increases the risk of root fracture. The most common type of fracture is usually a vertical one and the prognosis is poor (fig. 6). Extraction, or sectioning of the root in the case of a multi-rooted tooth, is all that can be recommended.

References

1 Hasler J F, Mitchell D F. Analysis of 1628 cases of odontalgia: A corroborative study. *J Indianapolis District Dent Soc* 1963; **17:** 23–25.
2 Mitchell D F, Tarplee R E. Painful pulpitis: a clinical and microscopic study. *Oral Surg* 1963; **13:** 1360–1370.
3 Seltzer S, Bender I B, Zionitz M. The dynamics of pulp inflammation: Correlation between diagnostic data and histologic findings in the pulp. *Oral Surg* 1963; **16:** 846–871: 969–977.
4 Garfunkel A, Sela J, Ulmansky M. Dental pulp pathosis; clinicopathologic correlations based on 109 cases. *Oral Surg* 1973; **35:** 110.
5 Dummer P H, Hicks R, Huws D. Clinical signs and symptoms in pulp disease. *Int Endod J* 1980; **13:** 27–35.
6 Moorer W R, Wesselink P R. Factors promoting the tissue dissolving capability of sodium hypochlorite. *Int Endod J* 1982; **15:** 187–196.
7 Walker R T. Emergency treatment—a review. *Int Endod J* 1984; **17:** 29–35.
8 Schein B, Schilder H. Endotoxin content in endodontically involved teeth. *J Endod* 1974; **1:** 19–21.
9 Sundquist G. Bacteriological studies of necrotic dental pulps. Umea University Odontological Dissertations, No. 7. Sweden, 1976.
10 Weine F S, Dewberry Jr J A. *Endodontic therapy*. 3rd ed. pp 8–15. St Louis: C V Mosby, 1982.
11 Watts A, Paterson R C. The response of the mechanically exposed pulp to prednisolone and triamcinolone acetonide. *Int Endod J* 1988; **21:** 9–16.

Tresiolan: Espe GmbH, Seefeld, Oberbayern, W. Germany.
Ledermix: Lederle Laboratories Division, Cyanamid of Great Britain Ltd, Gosport, Hampshire.
Cresatin: Merke, Sharp and Dohme, West Point, Pennsylvania 19486, USA.

5

Morphology of the Root Canal System

One factor that has influenced the success rate of endodontic treatment has been the improved understanding of root canal morphology. In 1925, when Hess and Zurcher first published their study,[1] it became clear that teeth had root canal systems rather than the simplified canals that had been previously described. An understanding of the architecture of the root canal system is an essential prerequisite for successful root canal treatment.

The pulp chamber consists of a single cavity with projections (pulp horns) into the cusps of the tooth (fig. 1). With age, there is a reduction in the size of the chamber due to the formation of secondary dentine, which can be either physiological or pathological in origin. Reparative or tertiary dentine may be formed as a response to pulpal irritation and is irregular and less uniform in structure.

The entrances (orifices) to the root canals are to be found on the floor of the pulp chamber, below the centre of the cusp tips. In cross-section, the canals are ovoid, having their greatest diameter at the orifice or just below it (fig. 2). In longitudinal section, the canals are broader buccolingually than in the mesiodistal plane. The canals taper towards the apex, following the external outline of the root. The narrowest part of the canal is to be found at the 'apical constriction', which then opens out as the apical foramen and exits to one side between 0·5 and 1·0 mm from the anatomical apex. Deposition of secondary cementum may place the apical foramen as much as 2·0 mm from the anatomical apex.

Lateral and accessory canals

The importance of lateral and accessory canals in pulpal and periodontal disease is still a subject of debate. Lateral canals form channels of communication between the main body of the root canal and the periodontal ligament space. They arise anywhere along its length, at right angles to the main canal. The term 'accessory' is usually reserved for the small canals found in the apical few millimetres and form the apical delta (fig. 3). Both lateral and accessory canals are formed due to a break in 'Hertwig's epithelial root sheath', or, during development, the sheath grows around existing blood vessels. Their significance lies in their relatively high prevalence.[2] Kramer found that the diameter of some lateral canals was often wider than the apical constriction.[3] Lateral canals are impossible to instrument and consequently sealing them is only moderately successful. Where two canals exist within the same root, for example the mesial root of a lower molar, lateral communication (anastomosis) often occurs between them.

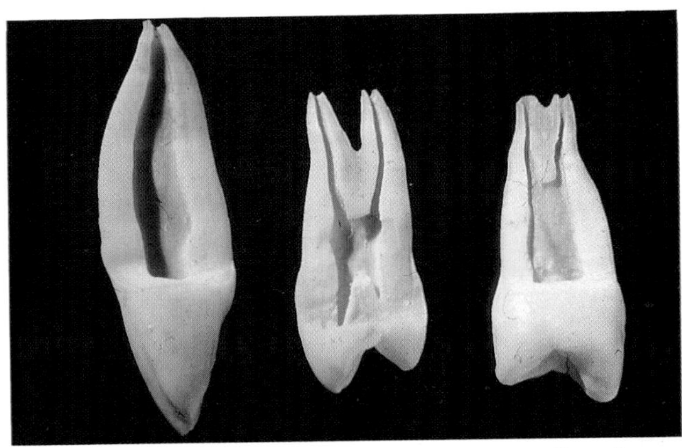

Fig. 1 Three examples of pulp chamber and canal configurations.

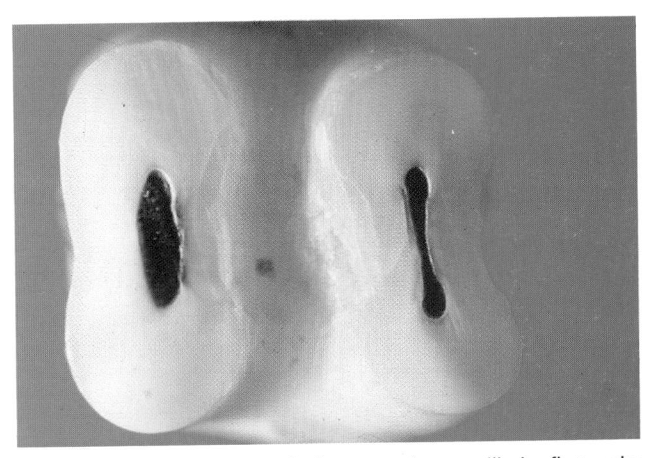

Fig. 2 Cross-section through the roots of a mandibular first molar, demonstrating the oval canals.

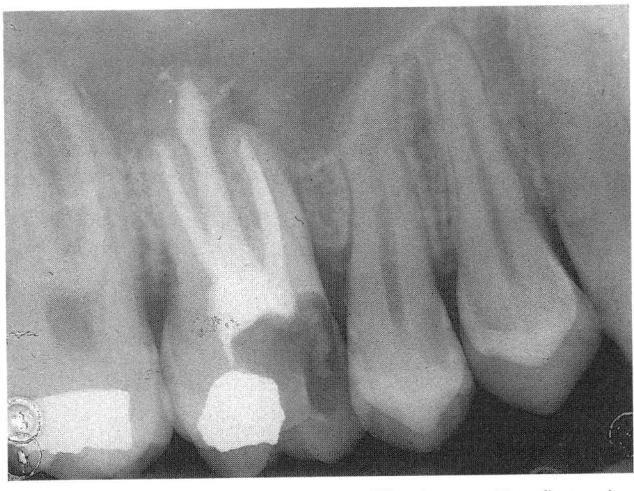

Fig. 3 Radiograph of a completed root filling in a maxillary first molar, demonstrating apical accessory canals in the palatal root.

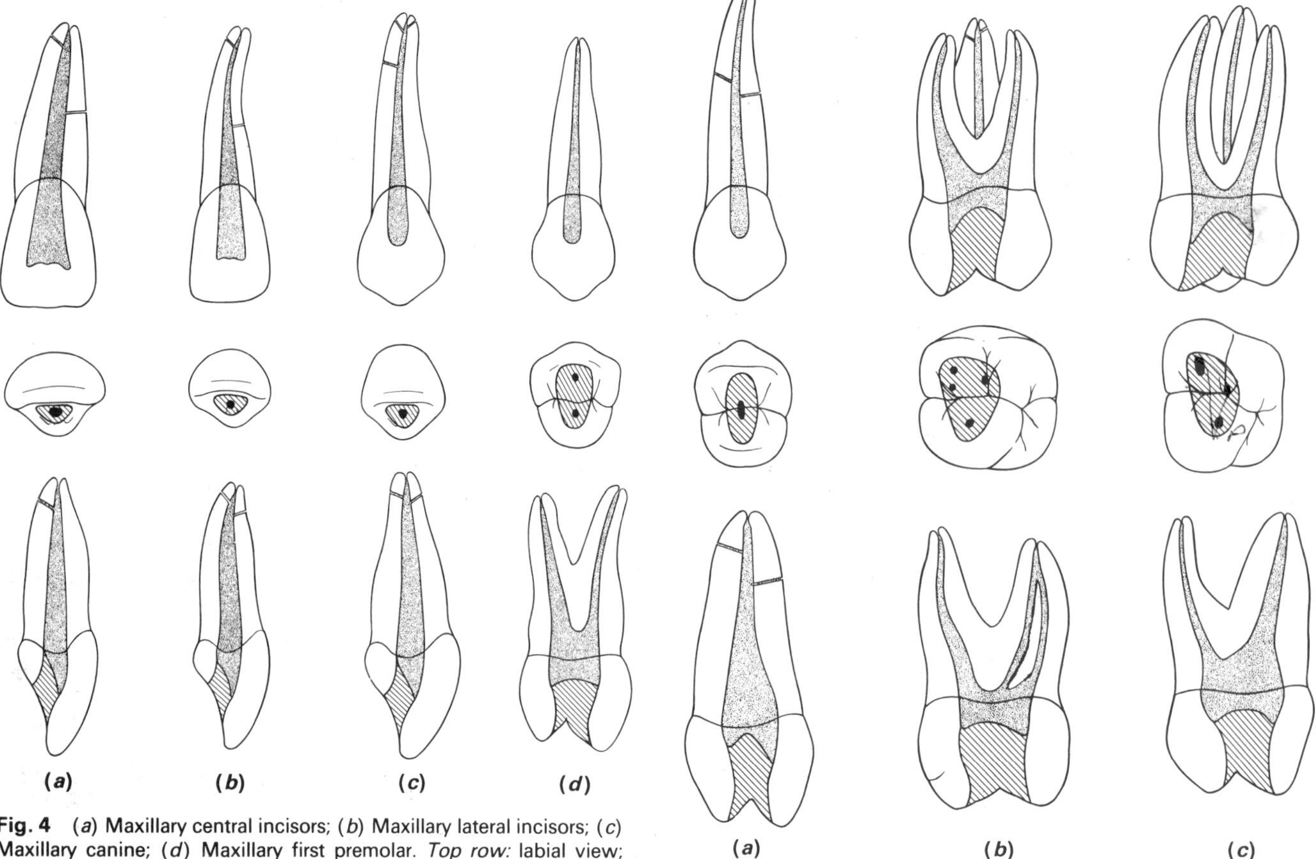

Fig. 4 (*a*) Maxillary central incisors; (*b*) Maxillary lateral incisors; (*c*) Maxillary canine; (*d*) Maxillary first premolar. *Top row:* labial view; *middle row:* occlusal view. *Bottom row:* mesiodistal view.

Fig. 5 (*a*) Maxillary second premolar; (*b*) Maxillary first molar; (*c*) Maxillary second molar. *Top row:* labial view; *middle row:* occlusal view. *Bottom row:* mesiodistal view.

Maxillary central incisors

These teeth almost always have one canal. When viewed on radiographs the canal appears to be fairly straight and tapering, but labiopalatally the canal will tend to curve either towards the labial or palatal aspect at about the apical third level. One feature to note is the slight narrowing of the lumen at the cervical level, which immediately opens up into the main body of the canal (fig. 4a).

Maxillary lateral incisor

Similar in shape to the central incisors, but fractionally shorter, the apical third tends to curve distally and the canal is often very fine with thin walls. Labiopalatally, the canal is similar to the central incisor, but there is often a narrowing of the canal at the apical third level. The root is more palatally placed, an important point when an apicectomy is carried out on this tooth (fig. 4b).

Maxillary canine

As well as being the longest tooth in the mouth, its oval canal often seems very spacious during instrumentation. However, there is usually a sudden narrowing at the apical 2–3 mm; this leads to a danger of overinstrumentation if too large a file is used at this level. The length of this tooth can be difficult to determine on radiographs, as the apex tends to curve labially and the tooth will appear to be shorter than it actually is (fig. 4c).

Maxillary first premolar

Typically, this tooth has two roots with two canals. In many ways this is the most difficult tooth to treat, as it can have a complex canal system. Variations range from one to three roots, but there are nearly always at least two canals present, even if they exit through a common apical foramen. The roots of these teeth are very delicate and at the apical third they may curve quite sharply buccally, palatally, mesially or distally, so instrumentation needs to be carried out with great care (fig. 4d).

Maxillary second premolar

In 40% of cases, this tooth, which is similar in length to the first premolar, has one root with a single canal. Two canals may be found in about 58% of cases.[4] The configuration of the two canals may vary with two separate canals and two exits, two canals and one common exit, one canal dividing and having two exits. In one study,[2] it was found that some 59·5% of maxillary second premolars had lateral canals. Like the first maxillary premolar, the apical third of the root may curve quite considerably, mainly to the distal, sometimes buccally (fig. 5a).

Maxillary first molar

This tooth has three roots. The palatal root is the longest, with an average length of 22 mm; the mesiobuccal and distobuccal roots are slightly shorter, at 21 mm average length. Pomeranz and Fishelberg found that *in vitro* 69% of these teeth had two canals in the mesiobuccal root, and 48% of these had separate foramina. The canals of the mesiobuccal root are often very fine and difficult to negotiate; consequently, more errors in instrumentation occur in this tooth than in almost any other. The curvature of the roots can be difficult to visualise from radiographs, and the second mesiobuccal canal is nearly always superimposed on the primary mesiobuccal canal. The palatal root has a tendency to curve towards the buccal and the apparent length on a radiograph will be shorter than its actual length (fig. 5b).

Maxillary second molar

This tooth is similar to the first maxillary molar, but slightly smaller and shorter, with straighter roots and thinner walls. Usually there are only three canals and the roots are sometimes fused (fig. 5c).

Maxillary third molar

The morphology of this tooth can vary considerably, ranging from a copy of the first or second maxillary molar to a canal system that is quite complex.

Mandibular central and lateral incisors

The morphology of these two teeth is very similar. The central incisor has an average length of 20·5 mm and the lateral is a little longer and has an average length of 21 mm. Over 40% of these teeth have two canals, but only just over 1% have two separate foramina. It is always necessary to check whether there is a second canal as, usually, only one canal can be detected on a radiograph (fig. 6a,b).

Mandibular canine

This tooth is similar to its opposite number, although not as long. On rare occasions, two roots may exist and this can cause difficulty with instrumentation (fig. 6c).

Mandibular first premolar

The canal configuration of this tooth can be quite complex. Vertucci[5] has shown that 25·5% of these teeth have two canals and two apical foramina. It is the way in which the second canal branches that can cause difficulty with instrumentation. Occasionally, the canal terminates with an extensive delta, making obturation of the accessory canals even more challenging (fig. 6d).

Mandibular second premolar

This tooth is similar to the first premolar, except that the incidence of a second canal is very much lower. One study[6] stated this to be 12%. Another study[5] revealed that only 2·5% had two apical foramina. Consequently, it is a much easier tooth to treat compared with the mandibular first premolar (fig. 7a).

Mandibular first molar

This is often the most heavily restored tooth in the adult dentition and seems to be a frequent candidate for endodontic treatment. Generally there are two roots and three canals: two canals in the mesial root and one large oval canal distally. According to Skidmore and Bjorndal,[7] one third of these molars have four canals. Occasionally three roots are to be found: two distal and one mesial, or one distal and two mesial. Anastomoses occur between the canals and accessory communication with the furcation area is a frequent finding. The mesiobuccal canal tends to exhibit the greatest degree of curvature (fig. 7b).

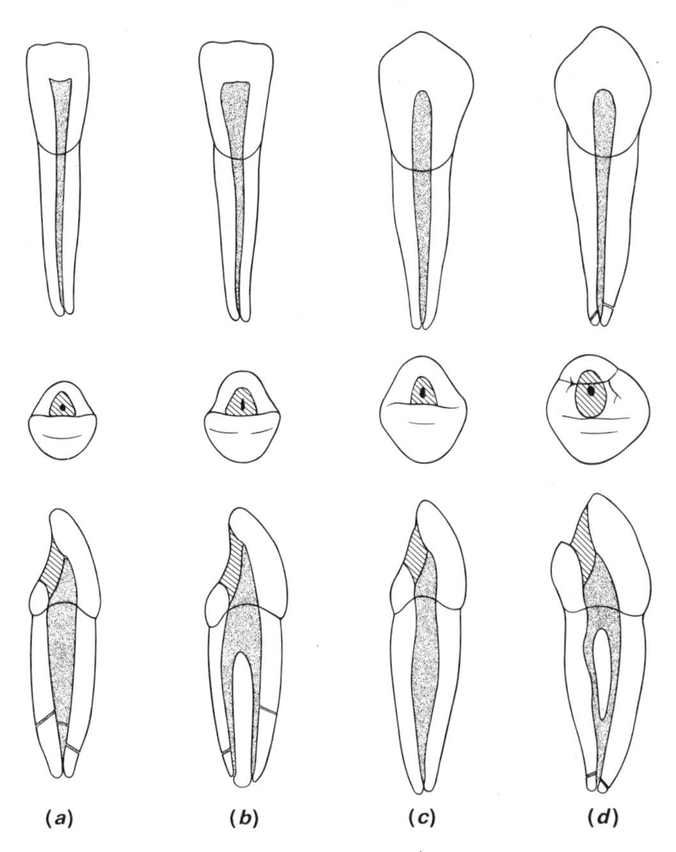

Fig. 6 (*a*) Mandibular central incisor; (*b*) Mandibular lateral incisor; (*c*) Mandibular canine; (*d*) Mandibular first premolar. *Top row:* labial view. *Middle row:* occlusal view. *Bottom row:* mesiodistal view.

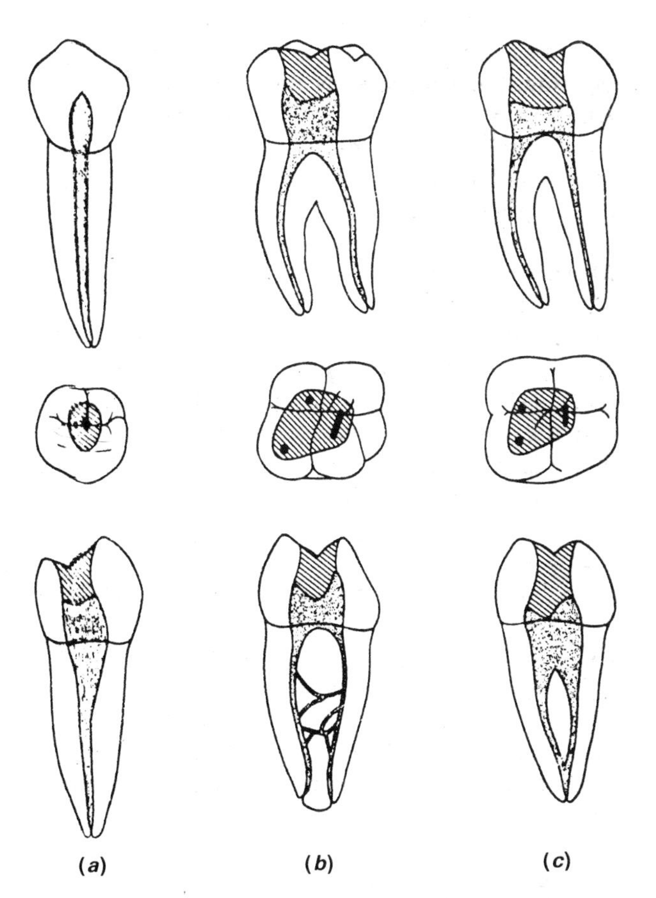

Fig. 7 (*a*) Mandibular second premolar; (*b*) Mandibular first molar; (*c*) Mandibular second molar.

Mandibular second molar

This tooth is similar to the mandibular first premolar, although a little more compact. The mesial canals tend to lie much closer together, and the incidence of two canals distally is much less. This tooth seems to be more susceptible to vertical fracture (fig. 7c).

Mandibular third molar

Together with the maxillary third molar, this tooth displays some of the most irregular canal configurations to be found in the adult dentition. The mesial inclination of the tooth generally makes access easier. The canal orifices are not too difficult to locate, but the degree of curvature of the apical half of the root canal system is often pronounced. Added to this, the apex is frequently poorly developed and lies close to the inferior alveolar canal.

Access cavity preparation

Unlike other aspects of dentistry, root canal treatment is carried out with little visual guidance; therefore, the difficulties that are likely to be encountered need to be considered. An assessment of the following features can be made after visual examination of the tooth, and study of a pre-operative periapical radiograph taken with a paralleling technique (fig. 8).

(1) The external morphology of the tooth.
(2) The architecture of the tooth's root canal system.

Table I Average lengths of permanent teeth

	Average lengths[8] (mm)	No. of canals	No. of roots	Reference no.
Maxillary anteriors				
Central incisor	23·70	1	1	
Lateral incisor	22·10	1	1	
Canine	27·30	1	1	
Maxillary premolars				
First premolar	22·30	1 (6·2%)	1	4
		2 (90·5%)	1–2	
		3 (1·1%)	2–3	
Second premolar	22·30	1 (40·3%)	1	4
		2 (58·6%)	1–2	
		3 (1·1%)	2–3	
Maxillary molars				
		(No. of canals in mesiobuccal root)		
First molar	22·30	1 (1 foramen) (38%)	3	9,10
		2 (1 foramen) (37%)		
		2 (2 foramina) (25%)		
Second molar	22·20	1 (1 foramen) (63%)	3	11
		2 (1 foramen) (13%)		
		2 (2 foramina) (24%)		
Mandibular anteriors				
Central incisor	21·80	1 (1 foramen) (58%)	1	12,13
Lateral incisor	23·30	2 (1 foramen) (40%)		
		2 (2 foramina) (1·3%)	1–2	
Canine	26·00	1 (94%)	1	13
		2 (2 foramina) (6%)		
Mandibular premolars				
First premolar	22·90	1 (1 foramen) (73·5%)	1	6
		2 (1 foramen) (6·5%)		
		2 (2 foramina) (19·5%)		
		3 (0·5%)		
Second premolar	22·30	1 (1 foramen) (85·5%)	1	
		2 (1 foramen) (1·5%)		
		2 (2 foramina) (11·5%)		
		3 (0·5%)		
Mandibular molars	**1–3**			
		(mesial canals)	1–3	
First molar	22·00	1 (1 foramen) (13%)		14
		2 (1 foramen) (49%)		
		2 (2 foramina) (38%)		
		(distal canals)		
		1		
		2 (28·9%)		7
		(mesial canals)	1–2	
Second molar	21·70	1 (1 foramen) (13%)		14
		2 (1 foramen) (49%)		
		2 (2 foramina) (38%)		
		(distal canals)		
		1 (1 foramen) (92%)		
		2 (1 foramen) (5%)		
		2 (2 foramina) (3%)		

(3) The number of canals present.
(4) The length, direction and degree of curvature of each canal.
(5) Any branching of the main canals.
(6) The relationship of the canal orifice to the pulp chamber and to the external surface of the tooth.
(7) The presence and location of any lateral canals.
(8) The position and size of the pulp chamber and its distance from the occlusal surface.
(9) Any related pathology.

Before commencement of root canal treatment the tooth must be prepared as follows:

(1) All caries and any defective restorations should be removed and made good. The tooth should be protected against fracture during treatment.
(2) The tooth should be capable of isolation.
(3) The periodontal status should be sound or capable of resolution.

It is crucial that the root canal does not become contaminated during instrumentation. If there is a danger of fracture, the cuspal height should be reduced to prevent this. If the loss of coronal tissue is extensive, there may be a need to restore the tooth temporarily with a temporary crown, copper ring or an orthodontic band. It is not always necessary to restore the tooth before carrying out endodontic procedures. Provided the tooth will anchor a rubber dam, the canals can be isolated from the oral cavity and a temporary seal placed over the canals, this will be sufficient (fig. 9). Even if the tooth will not hold a rubber dam, the teeth on either side of it may well do so (fig. 10).

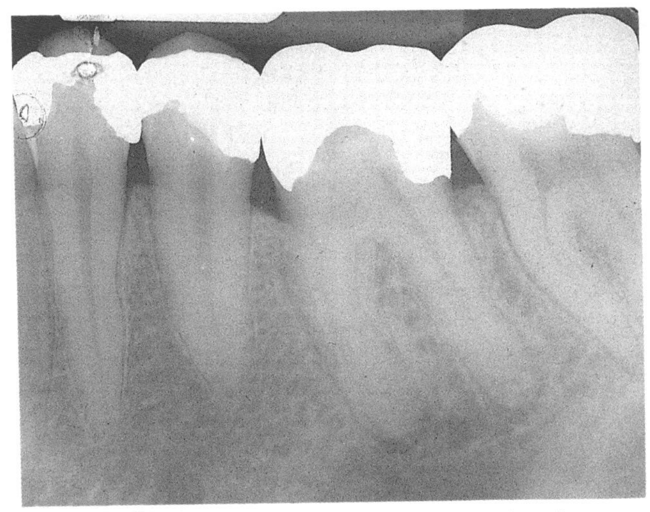

Fig. 8 All pre-operative radiographs must be taken using a long cone paralleling technique to give an image that corresponds with the true length of the tooth.

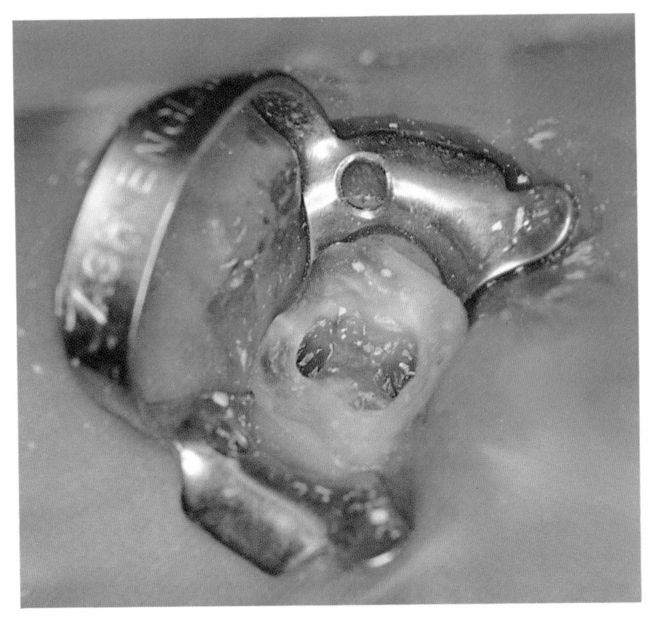

Fig. 9 Good visual access is essential. It is not necessary to restore the tooth before carrying out endodontic treatment, provided a rubber dam can be securely anchored to the tooth structure.

Rubber dam

The use of rubber dam is mandatory. On the rare occasion when it is impossible to place a rubber dam, the area must be isolated with cotton wool rolls, absorbent cheek guards and either gauze packs or butterfly sponges to protect the pharynx. In these circumstances, all the instruments used must be securely attached to a safety chain or dental floss. In this way, control of the operating environment may be gained, a prime requirement of good endodontic treatment.

Access

The objectives of access cavity preparation are to:

(1) Remove the entire roof of the pulp chamber so that the pulp chamber can be debrided.
(2) Enable the root canals to be located and instrumented by providing direct straight line access to the apical third of the root canals. (The initial access cavity may have to be modified during treatment to achieve this.)
(3) Enable a temporary seal to be placed securely in order to withstand any displacing forces.
(4) Conserve as much sound tooth tissue as possible and as is consistent with treatment objectives.

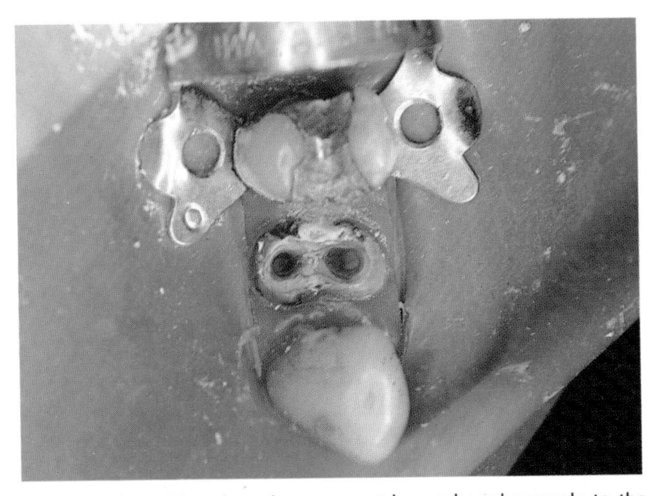

Fig. 10 If a rubber dam clamp cannot be anchored securely to the tooth, then the teeth on either side may well provide the anchorage.

The subsequent restoration of the tooth should always be considered first, and only the amount of coronal tissue sufficient for the successful completion of the endodontic

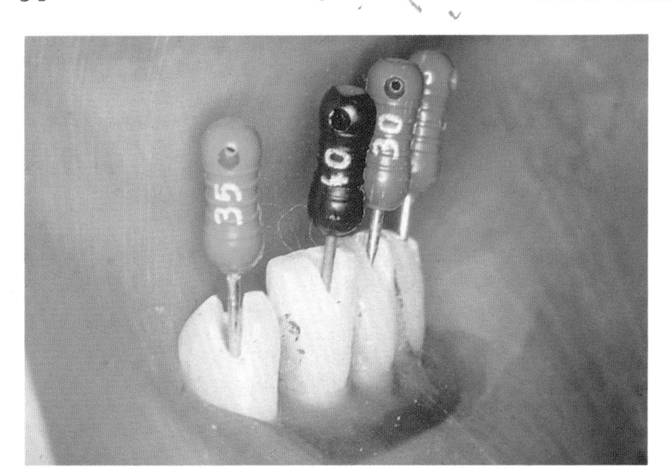

Fig. 11 Direct line access to the canals could only be gained via the incisal and labial surfaces.

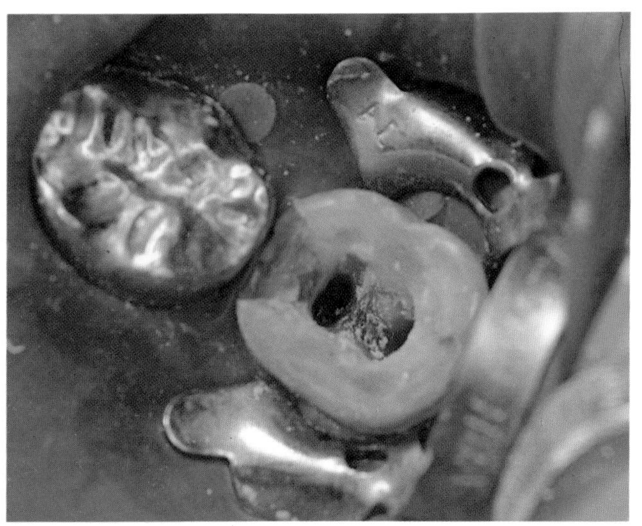

Fig. 12 Access to posterior teeth is often improved by either reducing the marginal ridge, or, in this instance, the mesial filling.

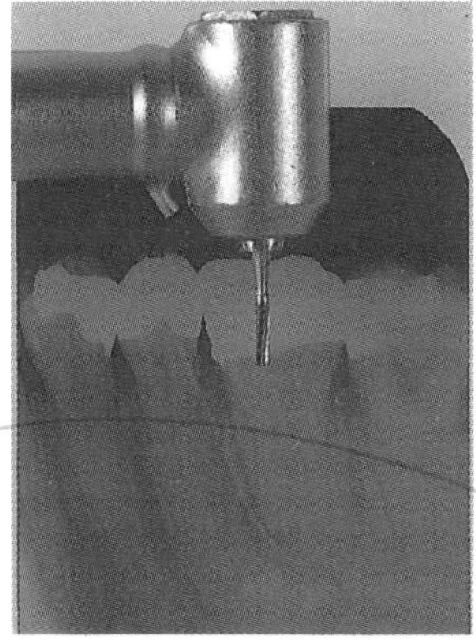

Fig. 13 Before commencing the access preparation, align the handpiece against the radiograph to check the depth and position of the pulp chamber.

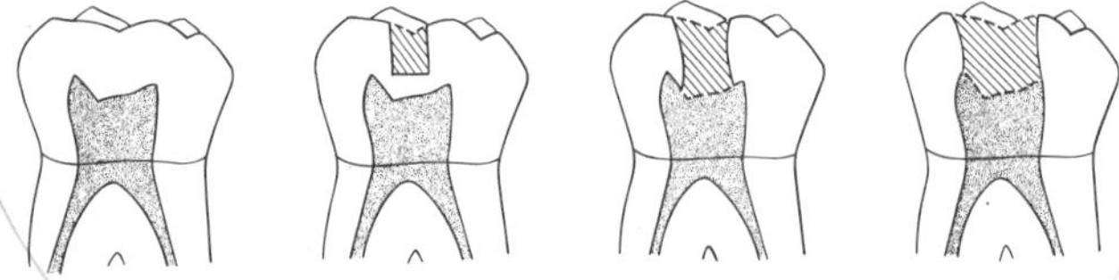

Fig. 14 Stages of access cavity preparation.

treatment should be removed. Sometimes it is necessary to approach the canal via the incisal edge, the buccal or palatal surfaces, as well as the occlusal surface to gain good access (fig. 11). If access to the back of the mouth is

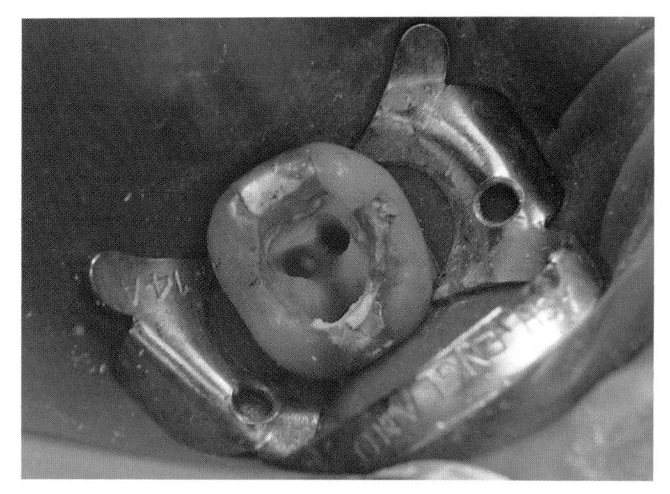

Fig. 15 Completed access preparation.

difficult, it is reasonable to consider reducing the marginal ridge of the tooth concerned to achieve this (fig. 12), or perhaps gain access through the mesiobuccal wall. Unless the root treatment is successful, any further restoration to the tooth will be put at risk.

Before beginning the access cavity preparation, it is wise to check the depth of the preparation by aligning the bur and handpiece against the radiograph, in order to note the position and depth of the pulp chamber in relation to the length of the bur in the handpiece (fig. 13).

The stages of access cavity preparation may be summarised as follows:

(1) The initial entry is made with a tungsten carbide bur in a turbine handpiece and the outline form completed as required (figs 4–7, middle rows). The final stages of removing the roof of the pulp chamber should be carried out at slow speed (fig. 14).
(2) The overhangs of dentine and enamel, especially the pulp horns, are removed using a long-shank round bur. The walls of the pulp chamber are then gently flared out towards the occlusal surface. The end result should be a gentle funnel-shape, with the larger diameter at the occlusal surface to provide resistance form for any temporary dressing.
(3) The contents of the pulp chamber are removed with a round bur at slow speed, taking care not to damage the floor of the pulp chamber. Any remaining pulp tissue and debris is cleared with an excavator from the floor of the pulp chamber and the canal orifices.
(4) The access cavity should then be flushed with a 2·5% solution of sodium hypochlorite to remove any residual debris.
(5) Locate the canal orifices with a DG 16 endodontic probe. Any alteration to the access outline form may now be undertaken to ensure a direct line of approach to the canal orifices.
(6) The cavo-surface angle can now be bevelled to facilitate access and to give extra support to the temporary filling.
(7) Finally, irrigate the entire access chamber with sodium hypochlorite again before commencing preparation of the root canal (fig. 15).

References

1 Hess W, Zurcher E. *The anatomy of the root canals of the teeth of the permanent dentition and the anatomy of the root canals of the deciduous dentition and first permanent molars*. London: Bale, Sons and Danielsson, 1925.
2 Vertucci F J, Seeling A, Gillis R. Root canal morphology of the human maxillary second premolar. *Oral Surg* 1974; **38:** 456–464.
3 Kramer I R H. The vascular architecture of the human dental pulp. *Arch Oral Biol* 1960; **2:** 177–189.
4 Bellizi R, Hartwell G. Radiographic evaluation of root canal anatomy of *in vivo* endodontically treated maxillary premolars. *J Endod* 1985; **11:** 37–39.
5 Vertucci F J. Root canal morphology of the mandibular premolars. *J Am Dent Assoc* 1978; **97:** 47–50.
6 Zillich R, Dowson J. Root canal morphology of the mandibular first and second premolars. *Oral Surg* 1973; **36:** 738–744.
7 Skidmore A E, Bjorndal A M. Root canal morphology of the human mandibular first molar. *Oral Surg* 1971; **32:** 778–784.
8 Bjorndal A M, Henderson W G, Skidmore A E, Kellner F H. Anatomic measurements of human teeth extracted from males between ages 17 and 21 years. *Oral Surg* 1974; **38:** 791–803.
9 Seidberg B H, Altman M, Guttoso J, Suson M. Frequency of two mesiobuccal root canals in maxillary first permanent molars. *J Am Dent Assoc* 1973; **87:** 852–856.
10 Pineda F. Roentgenographic investigation of the mesiobuccal root of the maxillary first molar. *Oral Surg* 1973; **36:** 253–260.
11 Pomeranz H H, Fishelberg G. The second mesiobuccal canal of maxillary molars. *J Am Dent Assoc* 1974; **88:** 119–124.
12 Benjamin K A, Dowson J. Incidence of two root canals in human mandibular incisor teeth. *Oral Surg* 1974; **38:** 122–126.
13 Vertucci F J. Root canal anatomy of the mandibular anterior teeth. *J Am Dent Assoc* 1974; **89:** 369–371.
14 Green D. Double canals in single roots. *Oral Surg* 1973; **35:** 689–696.

6

Preparing the Root Canal

Success in endodontic treatment depends largely on how well the root canal is cleaned and shaped. This chapter will cover the principles of root canal preparation, irrigation, root length determination, intra-canal medication, and temporary fillings. Two root canal preparation techniques will be described in detail, as the authors agree that, currently, they are the most efficient and suitable for clinical dental practice.

Principles and recent developments of root canal preparation

The principles of root canal preparation are to remove all organic debris and microorganisms and to shape the walls of the root canal so that the entire root canal space may be obturated. Currently, the root canal filling material of choice is gutta-percha, which requires a gradual even funnel-shaped preparation with the widest part coronally and the narrowest part 1·0 mm short of the root apex (fig. 1). Wide, relatively straight canals are simple to prepare, but fine curved canals can present considerable difficulties. A number of techniques have been described, all of which have been designed to produce a tapered preparation.

The standardised system

This technique was used for many years and requires each instrument, file or reamer, to be placed to the full working length. The canal was enlarged until clean white dentine shavings were seen on the apical few millimetres of the instrument. The filing was continued for a further 2 or 3 sizes, to complete the preparation. This method was satisfactory in straight canals, but was quite unsuitable for curved canals. As the instrument sizes increase, they become less flexible and cause errors in curved root canals. The common problems are ledging, zipping, elbow formation, perforation and loss of working length owing to impaction of dentine debris (fig. 2).

The stepback technique

The stepback technique was devised to overcome the problem of the curved root canal and has been described by Mullaney.[1] The apical region is first enlarged using files to a final master apical file size 25 or 30; each successively larger instrument is then inserted 1·0 mm less into the canal so that a taper is formed. In between placing each larger instrument, the master apical file is inserted to the working length to clear any debris collecting in the apical part of the canal; this is referred to as recapitulation.

The stepback technique helps to overcome the procedural errors (fig. 2) in slight to moderately curved canals, but in the more severely curved root canals problems still exist. There are three ways in which some of the problems of the curved root canal may be overcome, by using:

(1) a special filing technique;
(2) a file with a modified non-cutting tip;
(3) more flexible instruments.

Stepdown technique

This method was described by Goerig *et al.*[2] and has been followed by other, similar techniques such as the double flared[3] and the crown-down pressureless.[4] The principle of these techniques is that the coronal aspect of the root canal is widened and cleaned before the apical part (fig. 3). The obvious advantages of these methods over the stepdown are:

(1) It permits straighter access to the apical region of the root canal.

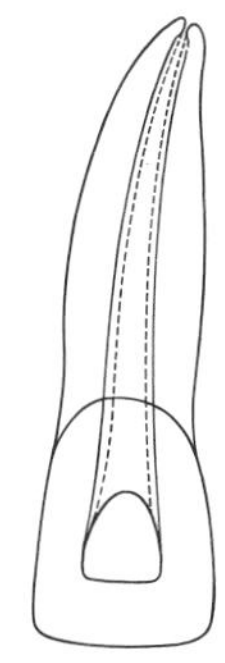

Fig. 1 The shape of the prepared root canal should be a gradual even taper, with the widest part coronally and the narrowest part 1·0 mm from the root apex.

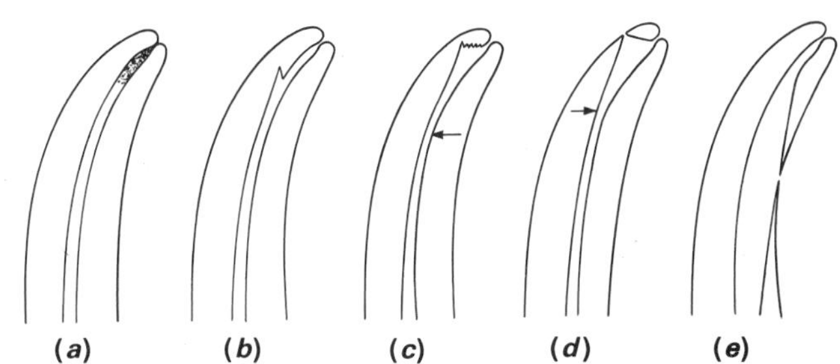

Fig. 2 Common errors in the preparation of curved canals. (*a*) Dentine debris packed into the apical part of the canal resulting in loss of working length. This may be prevented by recapitulation and copious irrigation. (*b*) Ledging due either to not precurving the instrument or forcing it into the canal. (*c*) Apical zip caused by not precurving the instrument or using a master apical file which is too large. Note the narrower part of the canal in (*c*) and (*d*) which is termed an elbow (*arrow*). This makes obturation of the root canal very difficult in the apical widened area. (*d*) Perforation due to persistent filing with too large an instrument. (*e*) Strip perforation caused by over preparing and straightening the curved canal.

(2) It eliminates dentinal interferences found in the coronal two-thirds of the canal, allowing apical instrumentation to be accomplished quickly and efficiently.
(3) The bulk of the pulp tissue debris and microorganisms are removed before apical instrumentation, which greatly reduces the risk of extruding material through the apical foramen and causing periapical inflammation. This should reduce the incidence of after-pain following preparation of the root canal.
(4) The enlargement of the coronal portion first allows better penetration of the irrigating solution.

The stepdown technique will be described later in this chapter.

The balanced force technique

The balanced force technique (Roane technique)[5] is another method of negotiating the curved root canal. Special flexible files are used with non-cutting tips. The file is inserted into the canal until slight resistance is felt and then rotated 90° clockwise to engage the dentine. Using light apical finger pressure the file is now rotated through 360° in an anticlockwise direction. It is this action that cuts dentine from the canal wall. The amount of apical pressure required to rotate the file anticlockwise is just sufficient to prevent it from winding out of the canal. The technique has a quite different feel from standard filing and must be practised. It is all too easy to fracture files if too much force is used. Using this method, curved canals may be prepared to the full working length, to a file size 55, without producing apical transportation.

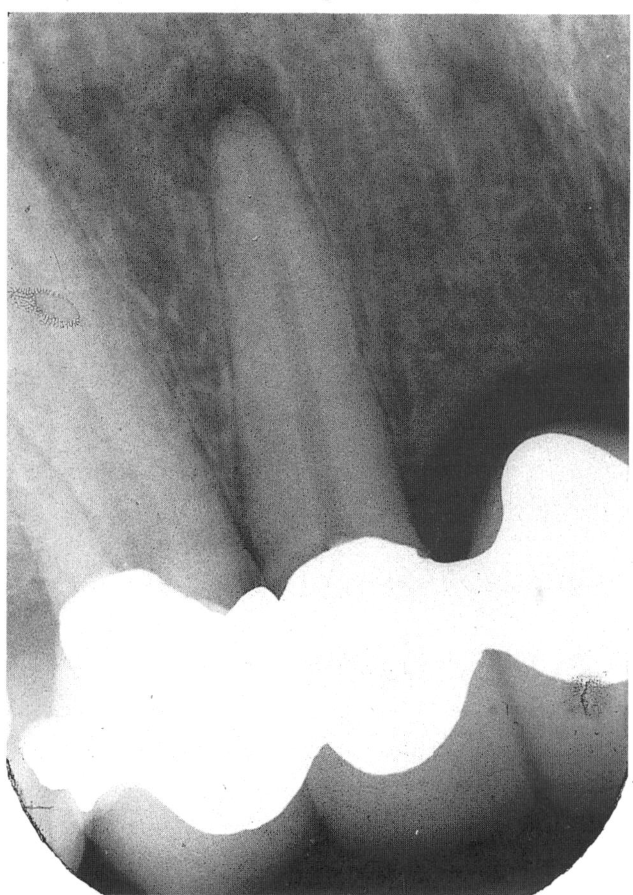

Fig. 3 The root canal in the ⌊2 (22) contains necrotic infected debris. The objective in endodontics is to remove the debris without extruding any through the apical foramen. It makes good sense to use a technique which cleans the coronal part first and then the apical portion.

Modified instrument tips

The design of the tip of the file has received much attention recently. There is evidence that a file which does not cut at its tip will produce less apical transportation than a standard tip (fig. 4).[6]

Anticurvature filing

This technique was put forward by Abou-Rass *et al.*[7] and supported by Lim and Stock.[8] Anticurvature filing is filing predominantly away from the inner curve of a root to reduce the risk of a strip perforation (fig. 5). The mesiobuccal roots of maxillary molars and the mesial roots of mandibular molars are the teeth most frequently at risk. The method is used only in canals with a moderate to severe curve. Figure 6 shows the cross-section of a canal and the clockwise circumferential movement of the file. The furcal wall is filed once, compared to the buccal mesial and lingual walls which are filed three times.

Flexible instruments

The manufacturers have responded to the call for more flexible instruments and there are many available. Alteration of the type of stainless steel has made it possible to produce flexible reamers and files. Changing the cross-sectional shape has also reduced the stiffness (Chapter 3). Recently, a new type of flexible instrument has been

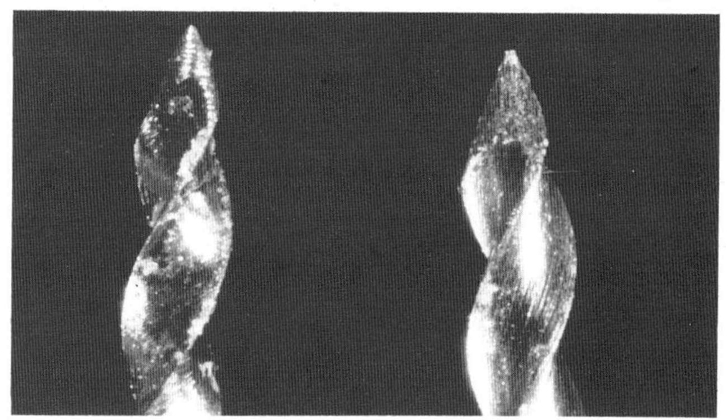

Fig. 4 File tip. (*left*) A standard tip of a file showing the cutting flutes; (*right*) A modified tip with no flutes.

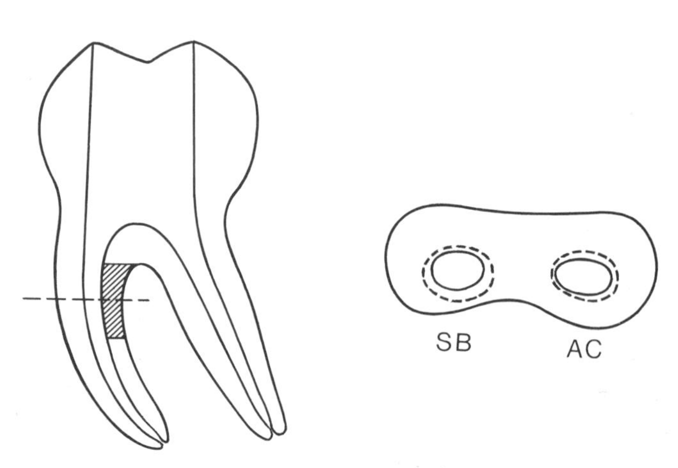

Fig. 5 Shaded area shows the danger zone where a strip perforation may occur. SB: stepback preparation shows risk of perforation; AC: anticurvature filing preserves a greater thickness of the furcal wall.

produced by Wildey and Senia,[9] which shows promise. The new instrument, termed a canal master,* incorporates three major features. First, it replaces the usual cutting tip with a non-cutting pilot; the pilot helps to limit transportation of the canal. Secondly, the length of the cutting area has been reduced from the standard 16 mm to 2–3 mm long. Finally, the diameter of the instrument's smooth round shaft remains constant and is reduced to increase its flexibility. Two versions of the canal master are available, one for hand and one for rotary instrumentation. The pilot is shorter for the hand instrument, being 0·75 mm as opposed to 2·0 mm for the rotary instruments.

Automated devices

There are now many automated handpieces on the market which claim to make the preparation of root canals quicker and more efficient. Evidence is unfortunately lacking to substantiate many of the claims made. Only with ultrasonic units is there some evidence that they may clean[10] and shape[11,12] root canals more efficiently than by hand, although other researchers[13] would disagree. Despite the lack of evidence, there is a growing trend towards the use of automated devices. The Sonic Air and the Pathfinder (Chapter 3, fig. 11) are particularly popular in Europe. There has been a rapid growth in the number of ultrasonic devices available.

Irrigation

The importance of irrigation in root canal preparation must be emphasised. A maxim in endodontics states that it is what you take out of a root canal that is important, not what you put in. Sodium hypochlorite is considered the most effective irrigant, as it is bactericidal, dissolves organic debris and is only a mild irritant.[14] The concentration recommended is approximately 2·5% available chlorine, which may be obtained by diluting Sainsbury's household bleach 1:1 with purified water BP. There are other commercially available sodium hypochlorite products, but it must be emphasised that there should be no other additives.

During preparation, the root canal should be kept wet, with copious irrigation used after each instrument. The irrigant in the canal is only replaced to the depth of insertion of the needle (fig. 7). The needle must remain loose in the canal while the irrigant is being injected, to prevent the solution passing into the periapical tissues (fig. 8). To obtain total replacement of irrigant solution in the root canal, the smallest needle available (30-gauge) should be placed at the apical foramen. Obviously, this is a most hazardous procedure and it is suggested that the irrigation needle is inserted to a maximum depth of 2·0 mm short of the working length. A file may then be worked in the apical 2·0 mm, to stir and withdraw the dentine debris further into the canal, so that it can be flushed away. There are several differently designed irrigation needle tips, but in the author's opinion these are of little importance compared to the diameter of the needle. Whatever the tip design, unless the needle can penetrate loosely, well into the root canal, irrigating, however copiously, will not remove dentinal debris.

Determination of root length

An estimate of the root length is made from the preoperative radiograph taken with a parallel technique. A file is placed carefully in the canal until it is within approximately 2·0 mm of the overall length. Before insertion the file

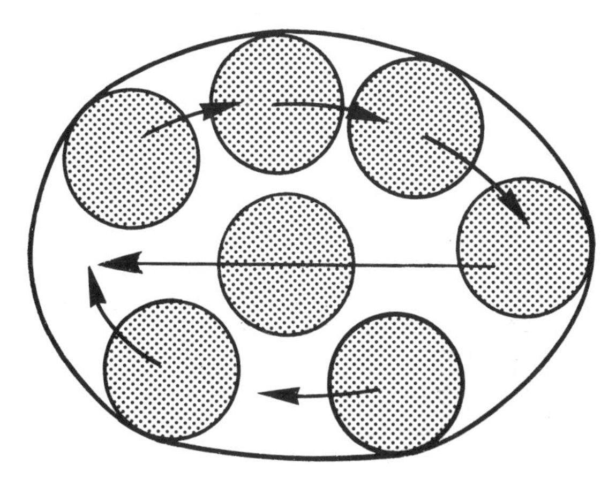

Fig. 6 Cross-section of a canal showing clockwise circumferential filing. The buccal, mesial and lingual walls are filed three times as much as the furcal wall.

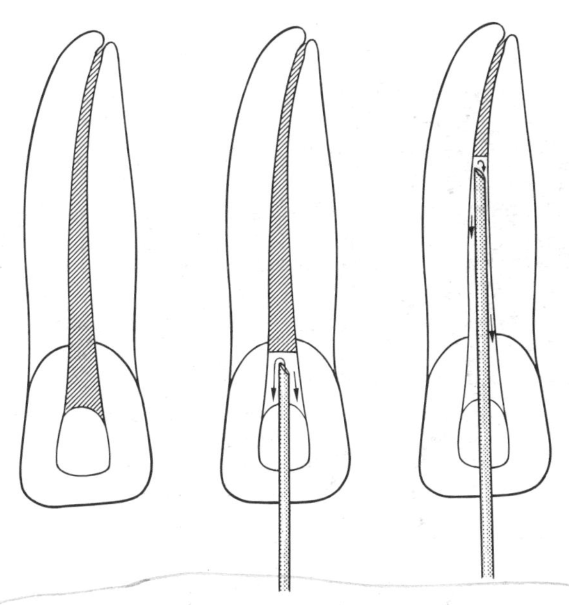

Fig. 7 The irrigant solution is only replaced in the root canal up to the depth of insertion of the needle.

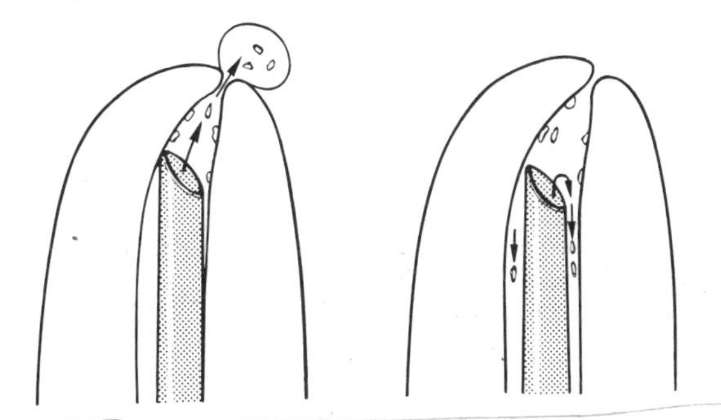

Fig. 8 The irrigation needle must be loose in the root canal to prevent expelling solution into the periapical tissues.

*Details can be found at the end of the chapter.

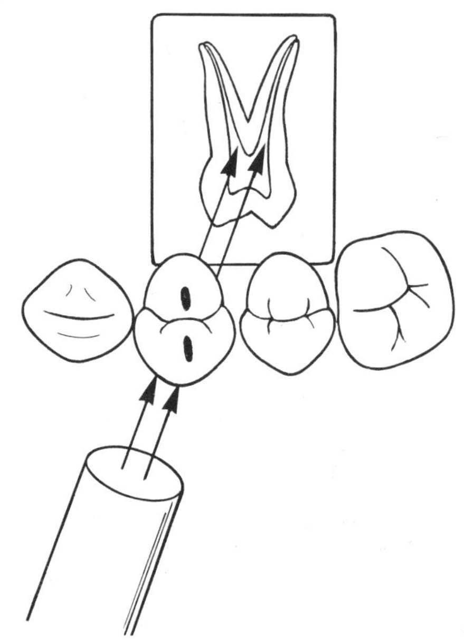

Fig. 9 The buccal object rule. When the x-ray cone is placed mesially and directed distally the buccal canal will appear the most distal on the radiograph.

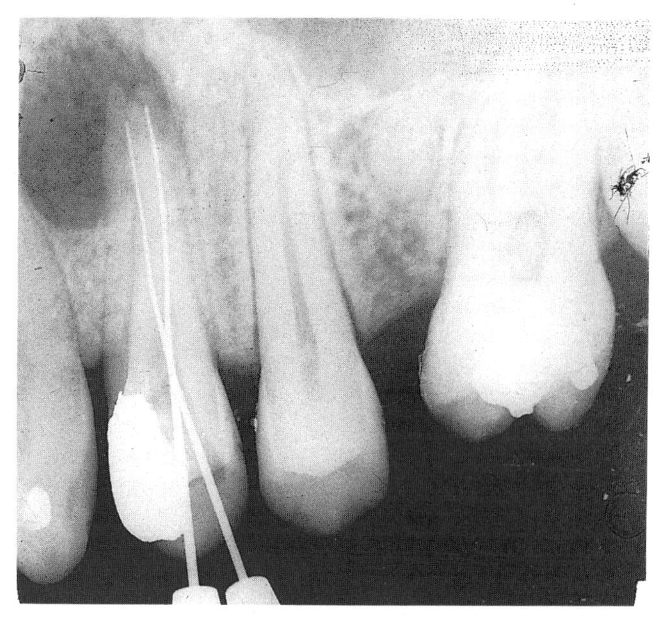

Fig. 10 The |4 (24) diagnostic radiograph was taken with the x-ray cone pointing distally. The buccal canal is the distal one on the radiograph and the mesial one the palatal canal.

should be precurved to the shape of the canal and gently manoeuvred into position. A silicone stop on the instrument shank is moved to a level opposite a reference point on the tooth. When taking diagnostic radiographs, use should be made of the 'buccal object rule', where there are two or more canals present in the root (figs 9 and 10). A second way of achieving the same result is to place a Hedstroem file in one canal and a K-file in the other, as the difference between the two is clear on the radiograph. The working length is assessed by measuring directly on the radiograph from the tip of the instrument to 1·0 mm short of the radiographic apex.

Electronic apex locators (Chapter 3) may be used in addition to radiography, to determine the working length. They are capable of accurate measurement, and, used in combination with radiography, will give the location of the apical foramen. They are essential when a patient is unable or unwilling to have radiographs taken. For routine use and when the outline of the canal is indistinct or the canal curves towards or away from the x-ray beam, they are most useful. The principle on which apex locators work is that they measure impedance of tissue which is a constant factor, plus the contents of the root canal to the apical foramen. Before a reading is taken, excess moisture is removed from the canal with one or two paper points. A file or reamer is inserted into the root canal and an electrical contact is made with the shank of the instrument. The device has a second electrode, which is placed in contact with the patient's oral mucosa. A digital display shows when the tip of the instrument reaches the apical foramen. There is wide agreement that these devices do not replace the radiograph, but are a most useful addition to the armamentarium. The level of sophistication of the devices is increasing, and most are now capable of measuring in a damp canal. There are several general rules concerning these apex locators:

(1) They do not work in a fluid containing dissociating ions. The presence in the root canal of frank pus, blood or sodium hypochlorite will give a false reading.
(2) The length from the apex, given in millimetres on the digital display, should be disregarded unless it is within 0·5 mm of the apex.
(3) Care must be taken to prevent an electrical contact between the instrument and a metallic restoration. The floor of the pulp chamber should be dried with cotton wool pledgets.
(4) There must be good contact between the oral electrode and the mucosa of the lip to prevent false readings.
(5) All the devices require some experience and 'feel' before the readings can be relied upon to be accurate. Small movements of the instrument near the correct length should be reproduced on the digital display.

Preparation of the root canal

Two techniques will be discussed: one uses hand instrumentation and the other ultrasound. The stepdown technique has been modified slightly from the original description by Goerig[2] in 1982, and the ultrasonic technique has been altered based on current research and to incorporate the concept of stepdown.

Stepdown technique

A pre-operative radiograph is taken, rubber dam placed and an access cavity cut. The preparation is divided into two parts: radicular access and apical instrumentation.

Radicular access

The pulp chamber is first copiously irrigated with sodium hypochlorite, followed by the sequential introduction of Hedstroem file sizes 15, 20 and 25 to a depth of 16–18 mm, or where the file binds against the canal walls (fig. 11). Each file is worked using circumferential filing and the canal irrigated. In curved canals, anticurvature filing is used.

This depth of 16–18 mm approximates the junction of the middle and apical thirds of the root. Where the roots

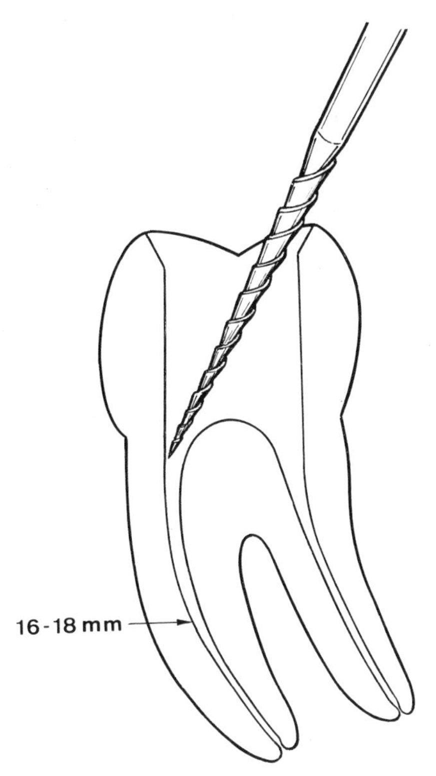

Fig. 11 Radicular access. Hedstroem file sizes 15, 20 and 25 are inserted sequentially to a depth of 16–18 mm or where the file binds against the walls.

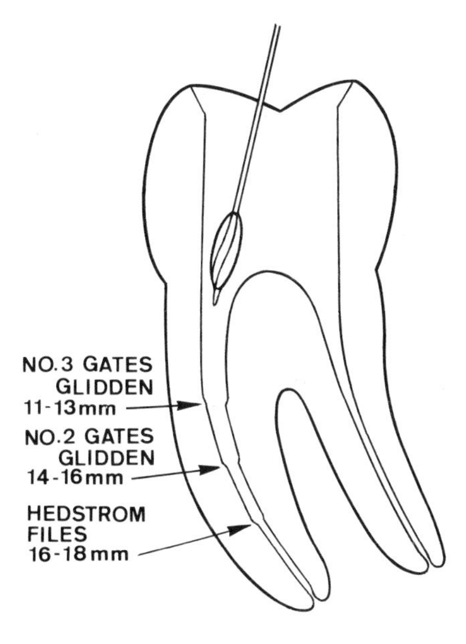

Fig. 12 Gates-Glidden burs sizes 2 and 3 are inserted into the root canal to the depths shown. The ledges in the diagram are for illustration purposes and should not occur in the root canal.

are short, the depth of insertion of the instruments is decreased. In small or calcified canals, the canal is first instrumented into the apical portion with sizes 08 and 10 K-files, to allow the placement of the size 15 Hedstroem file. In small canals, after filing with Hedstroem files, No. 10 K-file is used to renegotiate into the apical third. Ledging and blockage of the canal can be avoided by recapitulation.

Gates-Glidden burs are next introduced into the canal, beginning with a size 2 and followed by a size 3 (fig. 12). The No. 2 bur is inserted 14–16 mm into the canal from the occlusal reference point. The No. 3 bur is placed 11–13 mm into the canal and is directed apically and laterally away from the furcation. Gates-Glidden burs should be rotated with constant medium drill speed from the time they enter the canal until removed. If a bur does break, it usually does so near the handpiece head and may be retrieved easily from the tooth. Instrumentation with the stepdown technique in the radicular access is accomplished using only light pressure directed apically.

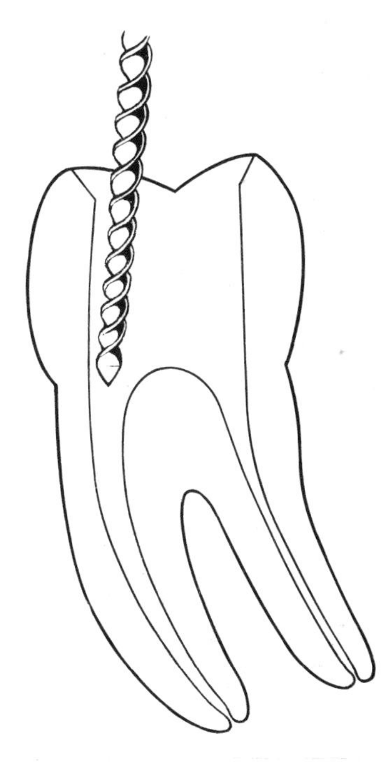

Fig. 13 Apical instrumentation is carried out using the stepback technique with files. Note the gradual even taper of the final preparation.

Apical instrumentation

The coronal flaring already carried out makes access to the apical portion of the root easier, as there are no dentinal obstructions and access is more direct.

The first step in apical instrumentation is to determine the working length, which is 1·0 mm short of the radiographic apex. This is carried out as described previously. Preparation of the apical part of the canal is achieved using the stepback technique. The choice of instrument depends on the operator's preference; the author uses K-flex files or Hedstroem files, although other types of flexible files would be suitable. The instruments should be precurved and only a push–pull motion is used, moving the file clockwise circumferentially. No turning motion is used.

Starting with a file size 15 at the working length, and progressing to sizes 20 and 25, an apical stop is made (fig. 13). Copious irrigation and recapitulation will prevent build-up of canal debris. The master apical file size is usually 25 or 30; larger instruments will start to produce ledging or zipping.

The apical portion is now tapered by stepping back. A file one size larger than the master apical file is placed 1·0 mm short of the working length and worked in the canal. Each successively larger file size is inserted 1·0 mm

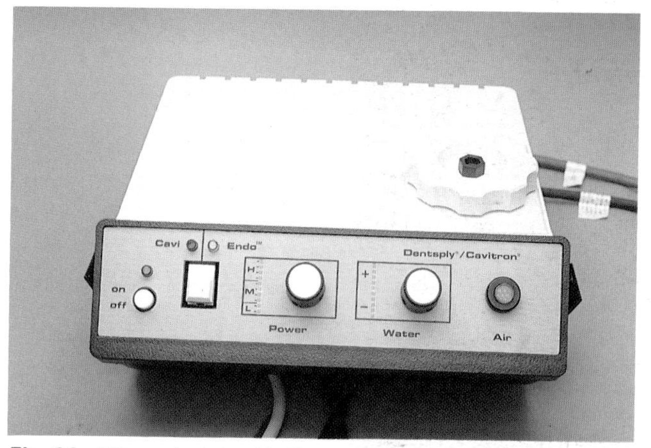

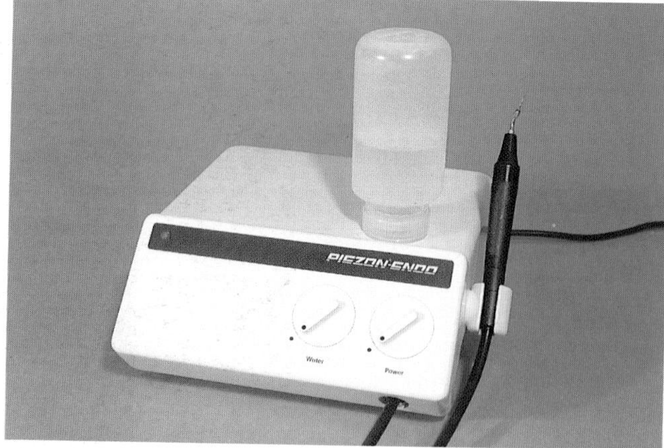

Fig. 14 Ultrasonic units. (*Left*) The Cavi-Endo unit manufactured by Dentsply is magnetostrictive. It is designed to use sodium hypochlorite and uses special K-files and diamond files. (*Right*) The Piezon Master manufactured by Electro Medical Systems (EMS) is a piezoelectric unit. It is also designed to accept sodium hypochlorite and uses special K-files.

Fig. 15 An ultrasonic handpiece showing the irrigant passing down the shank of the file.

less than the previous size until the radicular access preparation is reached. In between each larger file, the master apical file is inserted to the full working length and irrigation is used to remove all the debris. This final stepping back is achieved quickly as a taper has already been produced with the smaller instruments when placing the apical stop.

Ultrasonic technique

Endosonics has been developed largely by Martin and Cunningham during the last 15 years. There are many endosonic units available and they may be divided into two types, magnetostrictive and piezoelectric (fig. 14). The magnetostrictive stack is composed of a series of metal plates which are water cooled. The metal plates, surrounded by a magnetic field, expand and contract. The piezoelectric, a type of crystal which deforms rapidly in a magnetic field, does not require water cooling, which means that it is simpler in design and more portable.

Ultrasound consists of acoustic waves which have a frequency higher than can be perceived by a human ear. The acoustic energy is transmitted to the root canal instrument, which oscillates at 20–40 000 cycles per second, depending on which unit is used.

Research has shown that endosonics can be used to both clean and shape root canals, although how it does so is not yet fully understood. The cleaning is achieved by acoustic streaming and not, as originally thought, by cavitation.[15,16] Sodium hypochlorite is necessary[17] and a concentration of 2·5% available chlorine is recommended. Some of the ultrasonic units are not designed to accept sodium hypochlorite through the system and so, if water is used, they will be less efficient in the cleansing of root canals. Even when units designed to take sodium hypochlorite are used, daily maintenance must be carried out to prevent damage, particularly to metals, because the irrigant is corrosive.

The irrigant passes down the shank of the instrument and into the root canal, producing a continuous and most efficient system (fig. 15). Acoustic streaming is produced by the rapid file oscillations in the irrigant within the root canal. Figure 16 illustrates the severe turbulence created in the fluid.

The amount of pressure that should be used against the walls of the canal with the instrument is in dispute. Using light pressure the file will cut most efficiently at its tip, thereby producing apical widening and an elbow. If more pressure is employed, the tip will be constrained, which will severely limit the cutting power but improve the resultant shape. Using increased pressure constrains the tip but has less effect on the remainder of the flutes, so that more dentine is removed from the middle and coronal thirds of the canal. The author recommends that a moderate force of approximately 60 g is used. A recent study[18] found the average force used by nine operators on an endosonic unit was 30 g.

Fig. 16 Acoustic streaming. The photograph was taken of a file oscillating on the surface of water on which were floating fine spherical particles of acrylic. The pattern of movement of the particles is illustrated. There are numerous small vortices lying close to the file.

Ultrasonic preparation
A pre-operative radiograph is taken, rubber dam placed and an access cavity cut. A high-speed aspirator will evacuate the sodium hypochlorite welling out of the access cavity, but it is necessary to ensure a good seal with the rubber dam. Any minor gaps remaining around the tooth may be sealed using Cavit. The diagnostic radiograph is taken to determine the working length; in fine canals a size 08 or 10 instrument is used. The canal must be widened to a size 10 by hand. Using continuous irrigation with sodium hypochlorite, a number 15 ultrasonic K-file is gradually introduced into the root canal. The file is slowly worked down to 1·0 mm short of the working length; this takes about 1 minute. The file is slowly worked down to 1·0 mm short of the working length, taking about 1 minute, and then worked at that depth for a further minute. Two types of movement are used:

(1) Circumferential filing, similar to a hand-operated file, is employed. In curved canals, anticurvature filing is used.
(2) Circular filing motion. The file is moved around the circumference of the canal, keeping the depth the file is inserted constant. This second type of movement should be used sparingly, as it tends to produce an elbow.

The size 15 file is changed to a size 25 ultrasonic diamond file which is used in the straight part of the canal only. Because the irrigant is passing down the file shank, rubber stops cannot be used. Fine plastic tubing may be purchased and a small piece placed at the correct level on the shank; alternatively, indelible marking ink may be used. The diamond file is used for 2 minutes.

The apical stop is now placed at the working length using hand instruments. The master apical file size should be 25 or 30 in fine curved canals. Occasional difficulty may be experienced in inserting the 25 diamond file, in which case Gates-Glidden burs may be used to flare the coronal portion of the canal.

Finally, the 15 ultrasonic K-file is used to remove any remaining debris. A third file movement may be added to the two already described. The file is held stationary in the canal, trying not to touch the walls; this will prevent the file movements from being dampened and so allow the maximum acoustic streaming. Figure 17 shows examples of two molars which have been prepared using the ultrasonic technique and filled with laterally condensed gutta-percha.

Intracanal medication
There is universal agreement that no intracanal medicament yet exists that will sterilise the root canal. The importance of biomechanical preparation cannot be stressed too strongly. Most authorities would condemn the use of paraformaldehydes,[19,20] as they are toxic to the periapical tissues and may actually delay healing rather than promote it. There is some evidence that two intracanal medicaments have some benefit: calcium hydroxide (Chapter 9) and steroids. Steroids are readily obtainable in the form of triamcinalone (ledermix paste). The paste may be wiped on the canal wall using a file or paper point which

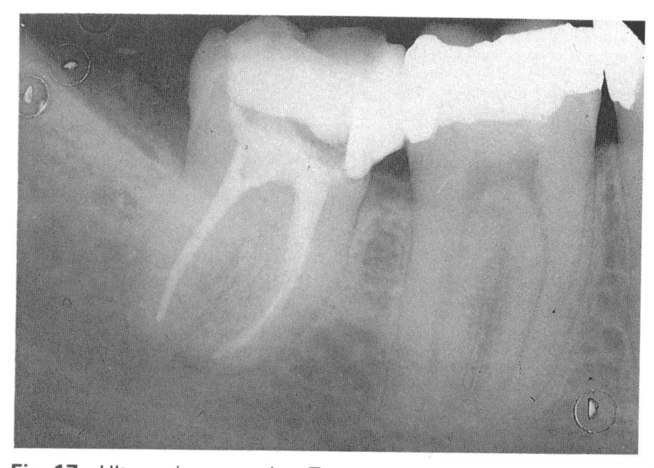

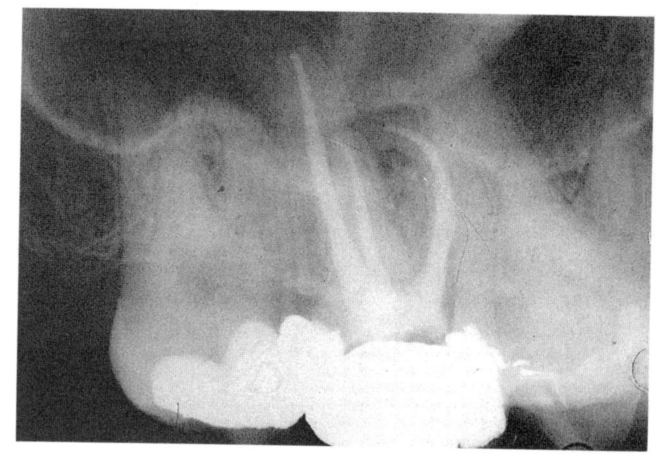

Fig. 17 Ultrasonic preparation. Two examples are shown of molars which have been prepared using the ultrasonic technique described and filled with laterally condensed gutta-percha.

is then withdrawn; a small amount is then placed on a pledget of cotton wool and sealed into the pulp chamber between appointments. An alternative is to mix the paste with a calcium hydroxide preparation and insert it on to the coronal part of the canals and pulp chamber and place a coronal seal. In the majority of cases, no medicaments are used. After preparation, the root canals are dried with paper points, sterile cotton wool is placed into the pulp chamber and the access cavity is sealed.

Temporary restorative materials

A temporary restorative material is required in endodontics as an interappointment dressing and until the tooth is ready to receive a permanent restoration. The material should prevent contamination of the root canal system and must be sufficiently strong to withstand the forces of mastication.

There are two temporary materials recommended, Cavit and IRM. Cavit provides a good seal, is simple to apply and quick to set. On the other hand, it lacks strength and will not stand up to masticatory forces. It should be confined to single surface fillings for periods not exceeding a week. Intermediate restorative material (IRM) does not provide a bacteria tight seal and is time-consuming to mix. It does, however, stand up well to forces of mastication.

It is a useful routine, as an interappointment dressing, to place a sterile pledget of cotton wool in the pulp chamber, followed by a layer of gutta-percha. The temporary restorative material is then placed over the gutta-percha. At the next visit, a high speed bur may be used to remove the IRM or Cavit without any danger of filling material lodging in the canal entrances. The gutta-percha provides a base for the restorative material and prevents the bur becoming caught in the cotton wool when the temporary filling is removed.

References

1 Mullaney T P. Instrumentation of finely curved canals. *Dent Clin N Am* 1979 **23:** 575–592.
2 Goerig A C, Michelich R J, Schult H H. Instrumentation of root canals in molars using the stepdown technique. *J Endod* 1982; **8:** 550–554.
3 Fava L R. The double flared technique: an alternative for biomechanical preparation. *J Endod* 1983; **9:** 76–80.
4 Morgan L F, Montgomery S. An evaluation of the crown-down pressureless technique. *J Endod* 1984; **10:** 491–498.
5 Roane J B, Sabala C L, Duncanson M G. The 'balanced force' concept for instrumentation of curved canals. *J Endod* 1985; **11:** 203–211.
6 Powell S E, Wong P D, Simon J H. A comparison of the effect of modified and non modified instrument tips on the apical canal configuration. Part II. *J Endod* 1988; **14:** 224–228.
7 Abou-Rass M, Frank A, Glick D. The anticurvature filing method to prepare the curved root canal. *J Am Dent Assoc* 1980; **101:** 792–794.
8 Lim S, Stock C. The risk of perforation in the curved canal: anticurvature filing compared with the stepback technique. *Int Endod J* 1987; **20:** 33–39.
9 Wildey W, Senia S. A new root canal instrument and instrumentation technique: a preliminary report. *Oral Surg* 1989; **67:** 198–207.
10 Haidet J, Reader A, Beck M, Meyers W. An *in vivo* comparison of the stepback technique versus a stepback/ultrasonic technique in human mandibular molars. *J Endod* 1989; **15:** 195–199.
11 Tang M, Stock C. An *in vitro* method for comparing the effects of different root canal preparation techniques on the shape of curved root canals. *Int Endod J* 1989; **22:** 49–54.
12 Tang M, Stock C. The effects of hand, sonic and ultrasonic instrumentation on the shape of curved root canals. *Int Endod J* 1989; **22:** 55–63.
13 Langeland K, Liao K, Pascou E. Work saving devices in endodontics: efficacy of sonic and ultrasonic techniques. *J Endod* 1985; **11:** 499–510.
14 The S, Maltha J, Plasschaert J. Reactions of guinea pig subcutaneous connective tissue following exposure to sodium hypochlorite. *Oral Surg* 1980; **49:** 160.
15 Walmsley D. Ultrasound and root canal treatment: the need for scientific evaluation. *Int Endod J* 1987; **20:** 105–111.
16 Ahmad M, Pitt Ford T, Crum L. Ultrasonic debridement of root canals: an insight into the mechanisms involved. *J Endod* 1987; **13:** 93–101.
17 Griffiths B, Stock C. The efficiency of irrigants in removing root canal debris when used with an ultrasonic preparation technique. *Int Endod J* 1986; **19:** 277–284.
18 Briggs PFA, Gulabiuala K, Stock CJR and Setchell DJ. The dentine removing characteristics of an ultrasonically energised K-file. *Int Endod J* 1989; **22:** 259–268.
19 Walton R. Intracanal medicaments. *DCNA Endodontics* 1984; 783–796.
20 Spangberg L. Intracanal medication. Ingle J and Taintor J. *Endodontics*. 3rd ed. pp 566–576. Philadelphia: Lea and Febiger, 1985.

Further reading

Endodontics. Ingle J and Taintor J F. 3rd ed. Access cavities and root canal preparation, pp 115–222. Philadelphia: Lea and Febiger, 1985 (see above).

Canal master: Brasseler USA Inc, Dental Rotary Instruments, 800 King George Boulevard, Savannah, Georgia 31419, USA.
Fine plastic tubing: Dentsply Ltd, Weybridge, Surrey.
Cavit: ESPE GmbH, Seefeld, Oberbayern, West Germany.
IRM: Dentsply Ltd, Weybridge, Surrey.

7

Filling of the Root Canal System

In current endodontic treatment the emphasis seems to be placed more on cleaning and preparing the root canal system than on filling it. This does not mean that root canal obturation is less important, but that its success depends on meticulous root canal preparation.

The purpose of a root canal filling is to seal the root canal system to prevent:

(1) microorganisms from entering and reinfecting the root canal system;
(2) tissue fluids from percolating back into the root canal system and providing a culture medium for any residual bacteria.[1]

In the past, attention has been focused on the importance of obtaining an hermetic apical seal. However, research has indicated that as well as sealing the root canal system apically, it is equally important to ensure that the coronal access to the canal itself has a fluid-tight seal, to prevent infection from the oral cavity.[2,3] Although numerous materials have been used to fill root canals, the most universally accepted is gutta-percha.

Properties of root canal filling materials[4]

Ideally, a root canal filling should be:

(1) Non-irritant.
(2) Dimensionally stable.
(3) Capable of sealing the canal laterally and apically, conforming with the various shapes and contours of the individual canal.
(4) Non-porous and impervious to moisture.
(5) Unaffected by tissue fluids and insoluble.
(6) Bacteriostatic.
(7) Radiopaque.
(8) Incapable of staining tooth or gingival tissues.
(9) Easily manipulated and should have ample working time.
(10) Easily removed from the canal if necessary.

Gutta-percha has a number of desirable properties. It is semi-solid and can be compressed and packed to fill the irregular shapes of a root canal using lateral or vertical condensation techniques. It is non-irritant and dimensionally stable. It will become plastic when heated or when used with solvents (xylol, chloroform, eucalyptus oil). It is radiopaque and inert, and can be removed from the canal when required for post preparation.

Its disadvantages are few. It is distorted by pressure and, consequently, can be forced through the apical foramen if too much pressure is used, and it is not rigid and so can be difficult to use in smaller sizes. Also, a sealer is necessary to fill in the spaces around the filling material.

Filling techniques

Several techniques have been developed for placing gutta-percha into the root canal system. Nevertheless, studies suggest that lateral condensation of gutta-percha is still the preferred technique.[5] However, as there is a demand for saving teeth with complex pathology and root canal morphology (figs 1 and 2), it is sometimes easier to combine the merits of various techniques in a hybrid form to simplify the filling procedure. Studies have shown that these are satisfactory, although not as easy as lateral condensation to carry out.[6,7]

Before a root filling is inserted, it is essential that the canals are dry. Any serous exudate from the periapical tissues indicates the presence of inflammation. If there is persistent seepage, calcium hydroxide may be used as a

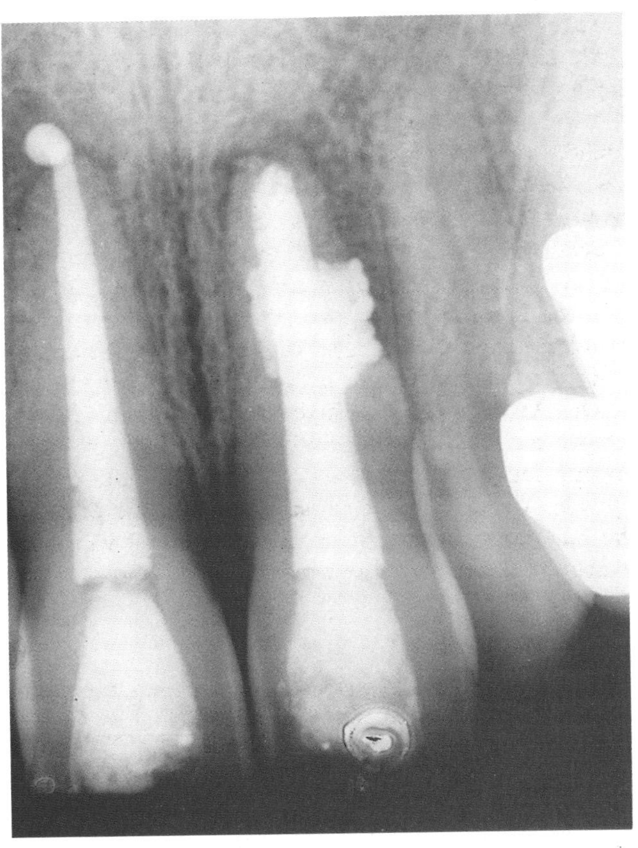

Fig. 1 Severe resorptive defects filled with high temperature gutta-percha technique (Obtura system).

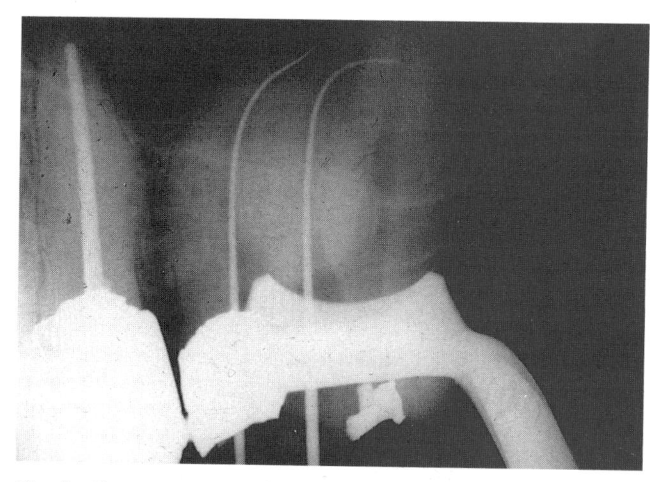

Fig. 2 Severe curvature of mesial and distal canals.

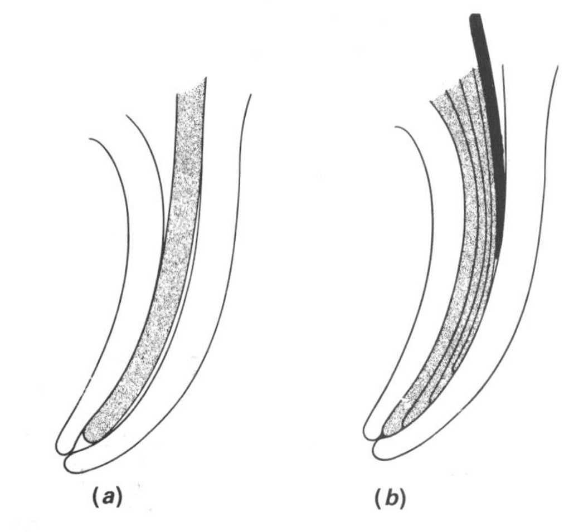

Fig. 3 (*a*) Master point seated within 1 mm of the working length. (*b*) Accessory points are added and condensed laterally until the entire canal space is sealed.

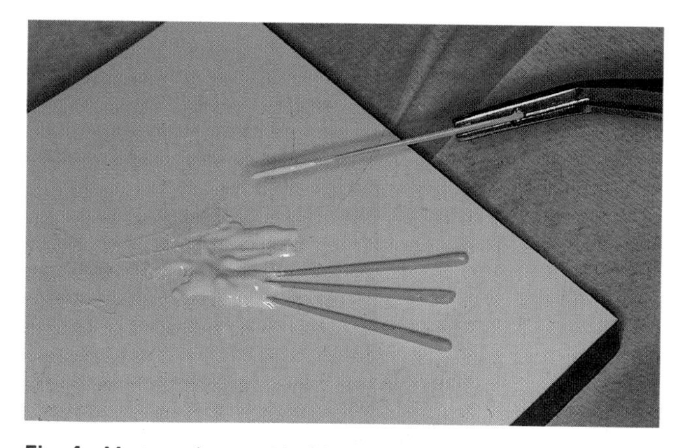

Fig. 4 Master point coated with sealer to paste canal walls. Accessory points ready for use.

root canal dressing until the next visit (calcium hydroxide BP mixed with local anaesthetic solution to a thick paste).

Lateral condensation of gutta-percha

The objective is to fill the canal with gutta-percha points (cones) by condensing them laterally against the sides of the canal walls. The technique requires selection of a master point, usually one size larger than the master apical file, which should seat about 1 mm short of the working length (fig. 3a). If the point reaches the working length, then either cut off 1 mm and try again, or use a larger size. Accessory points are then inserted alongside the master point and condensed laterally with a spreader until the canal is sealed (fig. 3b). There are two main types of spreading instruments for condensing gutta-percha: long-handled spreaders and finger spreaders. The main advantage of a finger spreader is that it is not possible to exert the high lateral pressure that might occur with long-handled spreaders. The chance of a root fracture is reduced and it is therefore a suitable instrument for beginners.

Procedure

(1) Coat the master point with sealer and paste the walls with the sealer before seating the point home into the canal (fig. 4).
(2) Place a spreader alongside the master point and condense using finger firm apical pressure only (figs 5 and 6). Leave the spreader *in situ* for 30 seconds. This is important as continuous pressure from the spreader is required to deform the gutta-percha point against the canal walls and to overcome its elasticity (fig. 7).
(3) Rotate the spreader a few times through an arc of 30–40° and withdraw.
(4) Select an accessory point and, after dipping its tip into sealer, place alongside the master point. Reinsert the spreader and laterally condense both points (fig. 8).
(5) Repeat sequence until canal is filled (figs 9 and 10).
(6) Remove excess gutta-percha from the canal orifice with a heated plugger, and firmly condense the remaining gutta-percha to seal the coronal access to the canal (fig. 11).
(7) Post preparation may be carried out at this stage if required.
(8) A periapical radiograph should be taken on completion, using a long cone parallel technique.

The completed root filling should be regularly reviewed and radiographs taken to check periapical state (fig. 12).

Lateral condensation of warm gutta-percha

A simple modification to the cold lateral condensation technique is to heat the spreader before insertion into the canal.[8] The softened material is easier to condense and will result in a denser root filling. Electrically heated spreaders are now available (fig. 13), but simply heating a spreader in a glass bead steriliser is effective (fig. 14).

Single gutta-percha point and sealer

The only advantage of a single gutta-percha point and sealer technique is its simplicity. The disadvantage is that the majority of sealers are soluble and the core material is unlikely to provide an effective seal.[8,9]

Vertical condensation of warm gutta-percha

Vertical condensation of warm gutta-percha is an excellent technique, but it is difficult and time-consuming, and extensive canal preparation is required.[8]

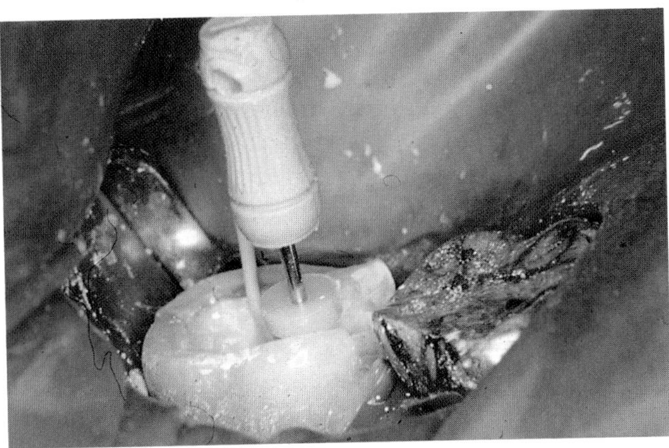

Fig. 5 Finger spreader inserted alongside master point and left *in situ* for 30 seconds.

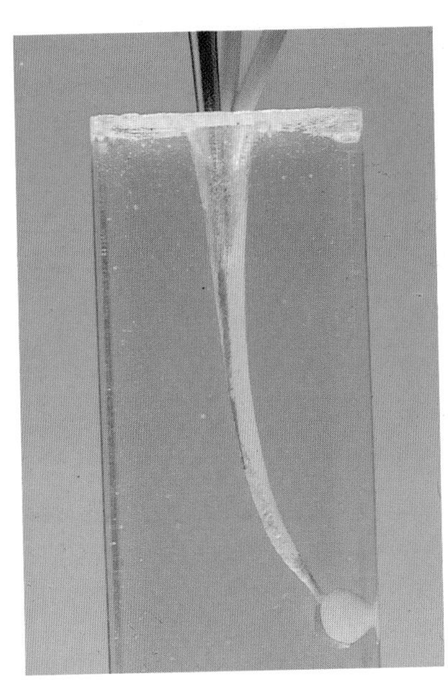

Fig. 8 Endo-vu block—Stage 3: Accessory point added and procedure repeated with spreader as before.

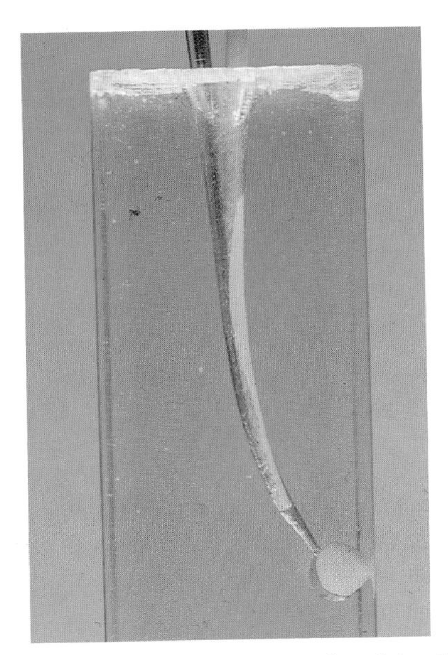

Fig. 6 Endo-vu block—Stage 1: Demonstration of the effect of lateral spreading forces. Note flow of gutta-percha around canal walls under pressure from spreader.

Fig. 9 Further accessory points added to canal.

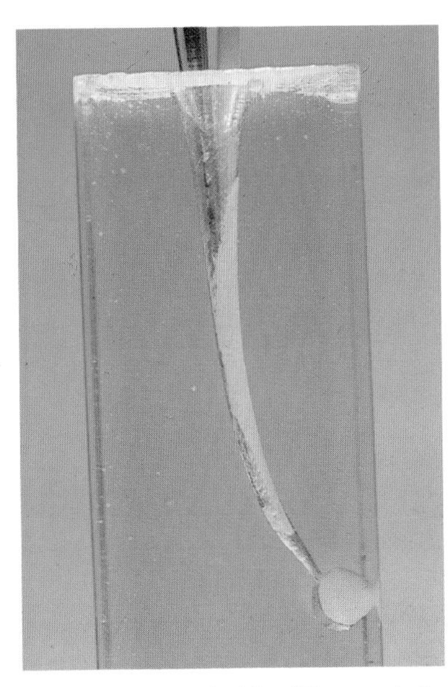

Fig. 7 Endo-vu block—Stage 2: After 30 seconds the gutta-percha has spread further round the canal walls providing tighter adaption.

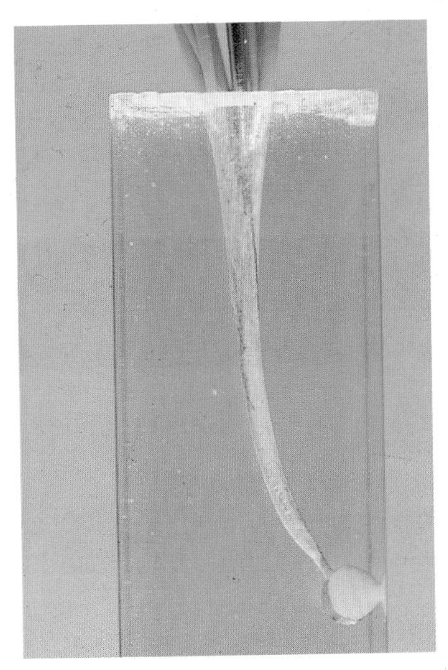

Fig. 10 Endo-vu block—Stage 4: As accessory points are added, the gutta-percha mass moves apically and laterally under pressure to seal the foramen.

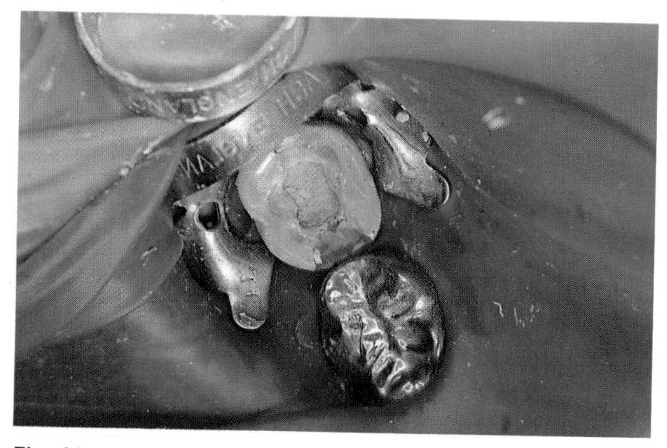

Fig. 11 Excess gutta-percha removed with heated plugger and the softened mass vertically condensed to seal the coronal access.

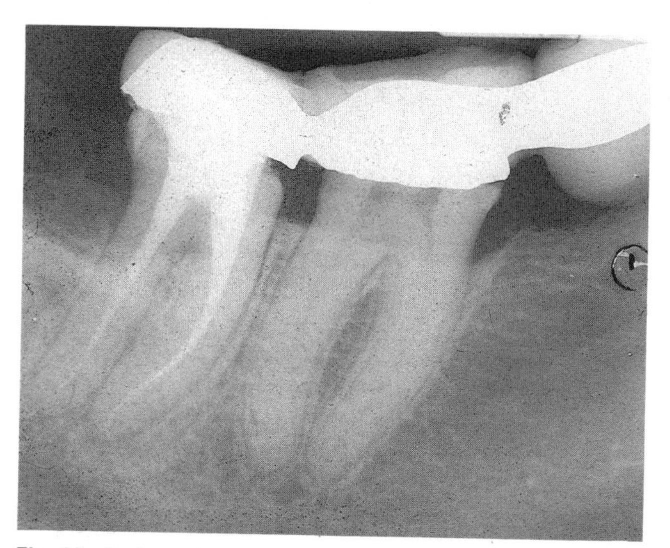

Fig. 12 Radiograph taken at completion.

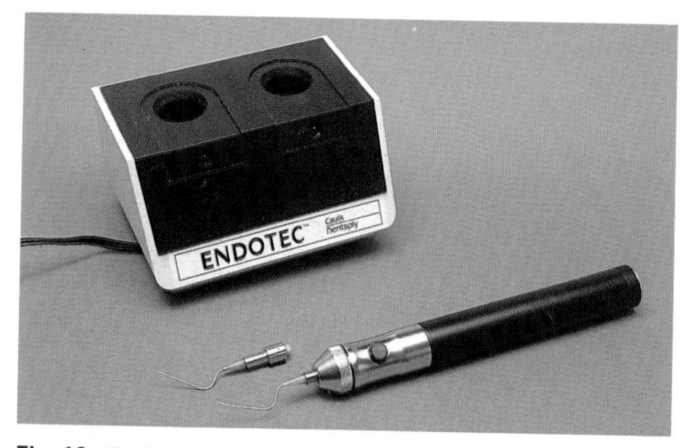

Fig. 13 Rechargeable battery spreaders.

Thermatic condensation of gutta-percha

McSpadden devised a handpiece-driven compactor which is effectively an inverted Unifile.[10] The frictional heat from the compactor plasticises the gutta-percha and the blades drive the softened material into the root canal under pressure. The main problem is lack of control over the apical portion of the gutta-percha, which is easily extruded through the apex in its softened state (figs 15, 16, 17, 18). In addition, there is a possibility of tissue damage caused by the heat generated from the compactor.[11]

Fig. 14 Glass bead steriliser provides a quick and easy method of heating spreaders.

Fig. 15 McSpadden compactor in demonstration block with master gutta-percha point only.

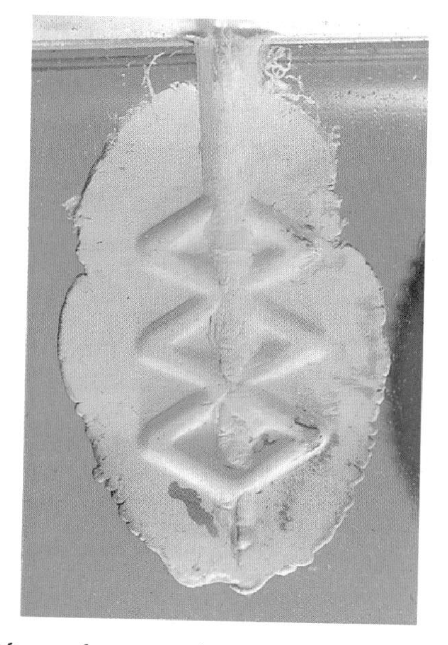

Fig. 16 After a few seconds rotation, the frictional heat has thoroughly plasticised the gutta-percha and the material has flowed between the block plates.

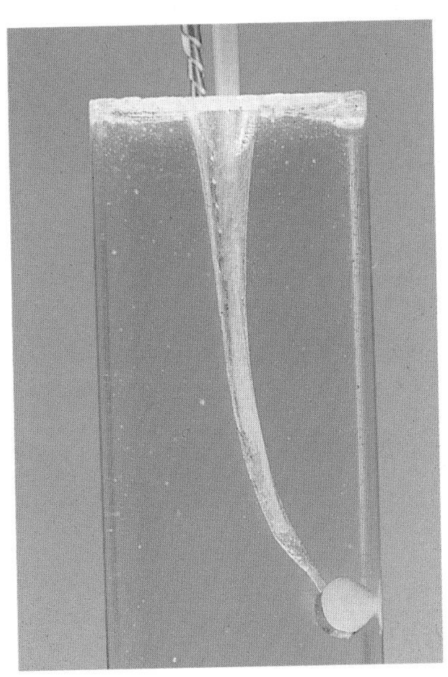

Fig. 17 Endo-vu block—Step 1: McSpadden compactor alongside master point. In curved canals the compactor will only reach the beginning of the canal curvature.

Fig. 18 Endo-vu block—Step 2: After compactor is activated a molten mixture of sealer and gutta-percha is formed which requires vertical compaction to offset any contraction on cooling.

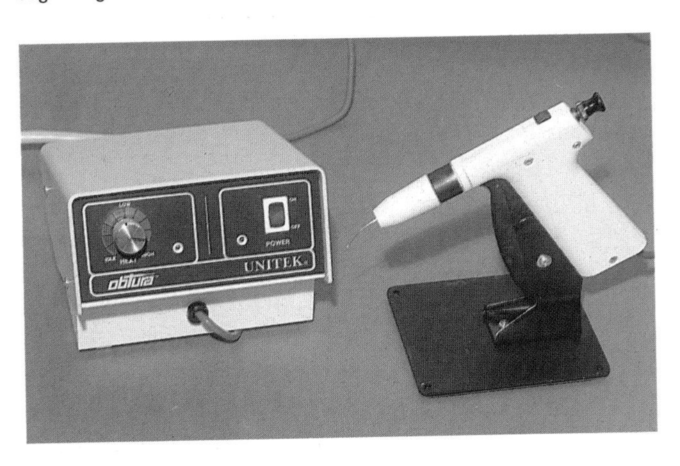

Fig. 19 High temperature injection moulded gutta-percha delivery system (Obtura).

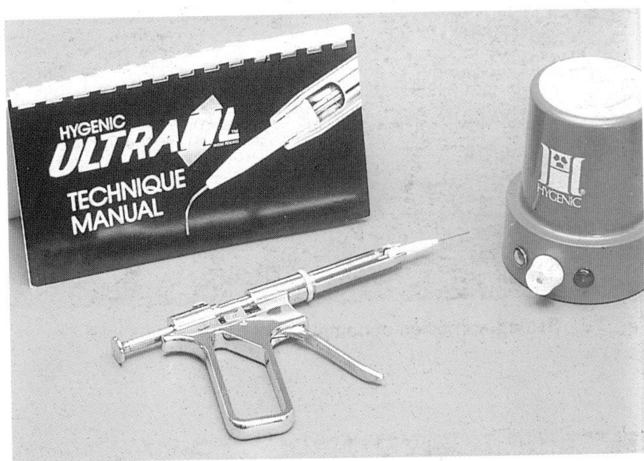

Fig. 20 Low temperature injection moulded gutta-percha delivery system (Ultrafil).

Injection-moulded thermoplasticised gutta-percha

Injection of a filling material to seal the canal system is an ideal shared by many. Two devices have been marketed which heat specially formulated gutta-percha for injection.[12] One system uses a high melting point gutta-percha, 'Obtura' (fig. 19). The other uses a low melting point system, 'Ultrafil' (fig. 20). Difficulty in controlling the flow of heated gutta-percha is a problem shared with mechanical thermocompaction techniques.[12] In addition, heated gutta-percha will contract on cooling and this may affect the quality of seal[13,14] (fig. 21).

Hybrid techniques

These have been developed to overcome difficulties that may be inherent in any one basic method.

Lateral condensation and thermocompaction of gutta-percha

Tagger described this technique as an adjunct to lateral condensation. Gutta-percha is first laterally condensed in the apical half of the canal, then a compactor is used to plasticise the gutta-percha in the coronal half of the canal (figs 22 and 23). The laterally condensed material in the apical half effectively prevents any apical extrusion and the

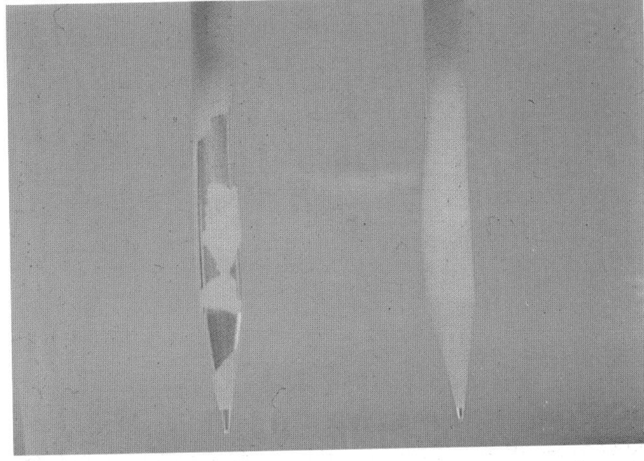

Fig. 21 Heated gutta-percha will contract on cooling away from the sides of the canal walls.

Fig. 22 Endo-vu block—Step 1: Hybrid technique. Lateral condensation and thermatic condensation of gutta-percha. After the gutta-percha has been laterally condensed at the apex, the compactor is introduced into the coronal third of the canal to plasticise the filling material.

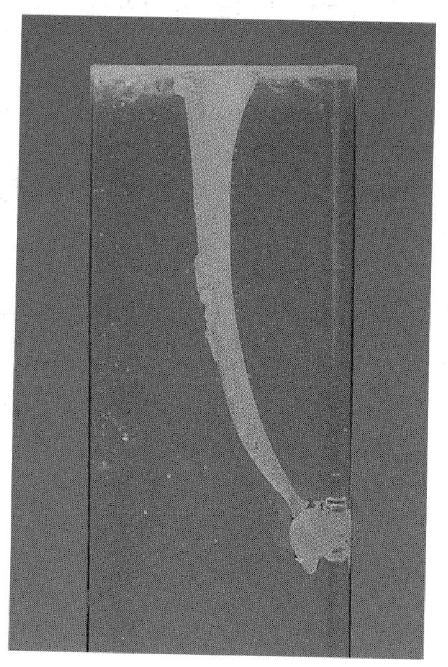

Fig. 23 Endo-vu block—Step 2: Hybrid technique. Lateral condensation and thermatic condensation of gutta-percha. Filling material has completely filled canal space. The fin that is apparent on the distal curvature, has been caused by the frictional heat from the compactor melting the resin. This demonstrates the considerable heat that can be generated and the compactor must be used for a second or two only.

Fig. 24 Endo-vu block—Step 3: Hybrid technique. Lateral condensation and thermatic condensation of gutta-percha. The spreader is introduced into the middle of the softened mass of gutta-percha and further accessory points introduced to offset any contraction.

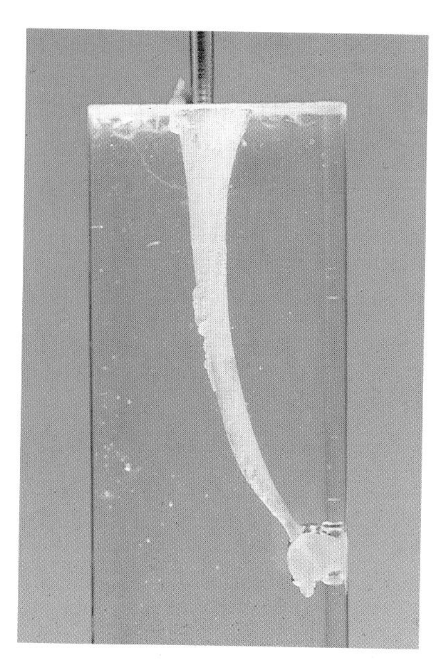

Fig. 25 Endo-vu block—Step 4: Hybrid technique. Lateral condensation and thermatic condensation of gutta-percha. Vertical compaction is carried out to seal the coronal access.

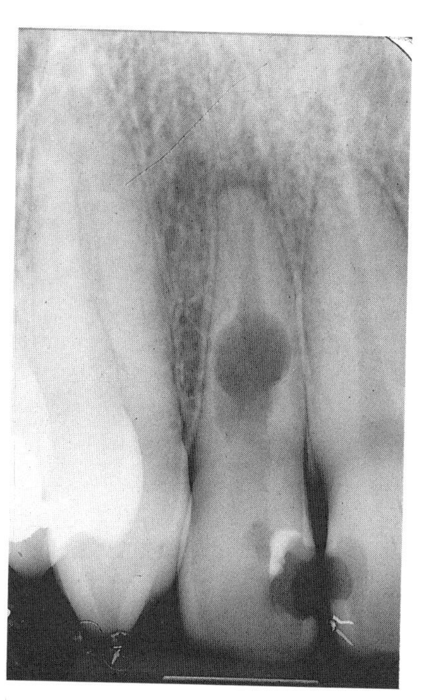

Fig. 26 Internal resorption in a maxillary lateral incisor.

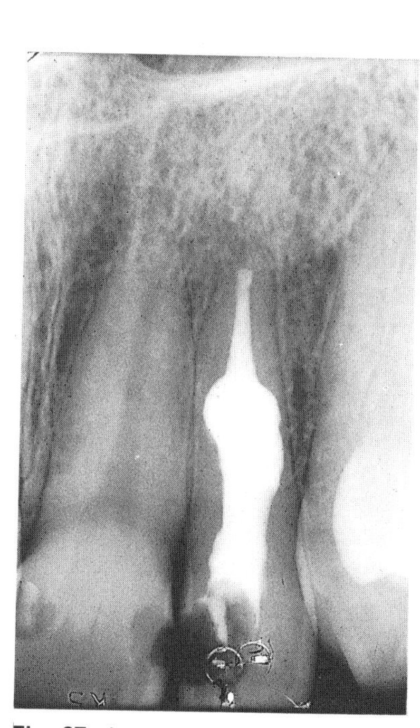

Fig. 27 Internal resorption in a maxillary lateral incisor. After treatment with calcium hydroxide, the canal was filled using high temperature, injection-moulded, thermoplasticised gutta-percha and lateral and vertical condensation technique to provide a stop apically.

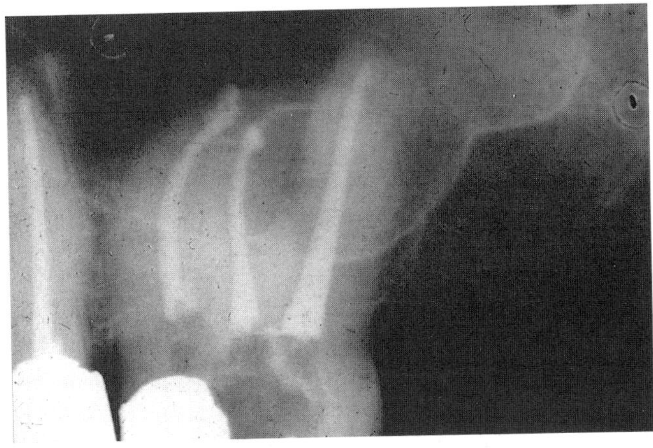

Fig. 28 Severe canal curvature filled using a low temperature, injection-moulded, thermoplasticised gutta-percha and lateral and condensation technique.

softened gutta-percha is thus forced against the dentine walls (figs 24 and 25).[7]

Lateral condensation and injection moulded, thermoplasticised gutta-percha

A similar approach can be adopted for presoftened gutta-percha, namely lateral condensation of gutta-percha in the apical half of the canal, followed by injection of thermoplasticised gutta-percha in the coronal half of the canal. The technique can be particularly useful in cases of internal resorptive lesions (figs 26, 27 and 28).

References

1 Harty F J. Filling the root canal. *Dent Update* 1977; **4:** 211–221.
2 Swanson K, Madison S. An evaluation of coronal microleakage in endodontically treated teeth. Part 1. Time periods. *J Endod* 1987; **13:** 56–59.
3 Swanson K, Madison S, Chiles S A. An evaluation of coronal microleakage in endodontically treated teeth. Part 2. Sealer types. *J Endod* 1987; **13:** 109–116.
4 British Endodontic Society. Guidelines for root canal treatment. *Int Endod J* 1982; **16:** 192–195.
5 Beer R, Gangler P, Rupprecht B. Investigation of the canal space occupied by gutta-percha following lateral condensation and thermomechanical condensation. *Int Endod J* 1987; **20:** 271–275.
6 Saunders E M. The effect of variation in thermomechanical compaction techniques upon the quality of the apical seal. *Int Endod J* 1989; **22:** 163–168.
7 Tagger M, Tamse A, Katz A, Korzen B H. Evaluation of the apical seal produced by a hybrid root canal filling method, combining lateral condensation and thermatic compaction. *J Endod* 1984; **10:** 299–303.
8 Harty F J. *Endodontics in clinical practice.* 2nd ed. pp 157–167. Bristol: John Wright and Sons, 1982.
9 Beatty R G, Vertucci F J, Zakariasen K L. Apical sealing efficacy of endodontic obturation techniques. *Int Endod J* 1986; **19:** 237–241.
10 McSpadden J T. Ash McSpadden compactor. Self study course for the thermatic condensation of gutta-percha. London: Dentsply, 1980.
11 Hardie E M. Further studies on heat generation during obturation techniques involving thermally softened gutta-percha. *Int Endod J* 1987; **20:** 122–127.
12 Lacombe J S, Campbell A D, Hicks M L, Pelleu G B. A comparison of the apical seal produced by two thermoplasticized injectable gutta-percha techniques. *J Endod* 1988; **14:** 445–450.
13 Bradshaw G B, Hall A, Edmunds D H. The sealing ability of injection-moulded thermoplasticized gutta-percha. *Int Endod J* 1989; **22:** 17–20.
14 Kersten H W. Evaluation of three thermoplasticized gutta-percha filling techniques using a leakage model *in vitro*. *Int Endod J* 1988; **21:** 353–363.

8

Calcium Hydroxide, Root Resorption, Perio-endo Lesions

Calcium hydroxide

Calcium hydroxide was introduced originally by Herman[1] in 1930 as a pulp-capping agent but its uses today are widespread in the field of endodontic therapy.

Mode of action

Despite extensive research, the mode of action of calcium hydroxide is still not fully understood. A calcified barrier may be induced when calcium hydroxide is used as a pulp-capping agent or placed in the root canal in contact with healthy pulpal or periodontal tissue. However, the calcium ions that form the barrier are derived entirely from the bloodstream and not from the calcium hydroxide.[2] The hydroxyl group is considered to be the most important component of calcium hydroxide as it provides an alkaline environment which encourages repair and active calcification. The alkaline pH induced not only neutralises lactic acid from the osteoclasts, thus preventing a dissolution of the mineral components of dentine, but could also activate alkaline phosphatases which it is thought play an important role in hard tissue formation. The calcified material which is produced appears to be the product of both odontoblasts and connective tissue cells and may be termed osteodentine. The barrier, which is composed of osteodentine, is not always complete and is porous (fig. 1).

In external resorption, the cementum layer is lost from a portion of the root surface, which allows communication through the dentinal tubules between the root canal and the periodontal tissues. It has been shown that calcium hydroxide expands to 2·5 times its initial volume on contact with water and so can diffuse through dentinal tubules. Tronstad[3] demonstrated that untreated teeth with pulpal necrosis had a pH of 6·0 to 7·4 in the pulp dentine and periodontal ligament, whereas, after calcium hydroxide had been placed in the canals, the teeth showed a pH range in the peripheral dentine of 7·4 to 9·6.

Tronstad[3] suggests that calcium hydroxide may have other actions; these include, for example, arresting inflammatory root resorption and stimulation of healing. It also has a bactericidal effect and will denature proteins found in the root canal, thereby making them less toxic. Finally, calcium ions are an integral part of the immunological reaction and may activate the calcium-dependent adenosine triphosphatase reaction associated with hard tissue formation.

Presentation

Calcium hydroxide can be applied as a cement, in paste form or powder and liquid, depending on the treatment. Different proprietary brands of the paste form are available commercially (fig. 2), although ordinary calcium hydroxide BP powder may be obtained from a chemist and mixed with barium sulphate powder to provide radiopacity in the ratio 8:1, respectively. Because of the antibacterial effect of calcium hydroxide, it is not neces-

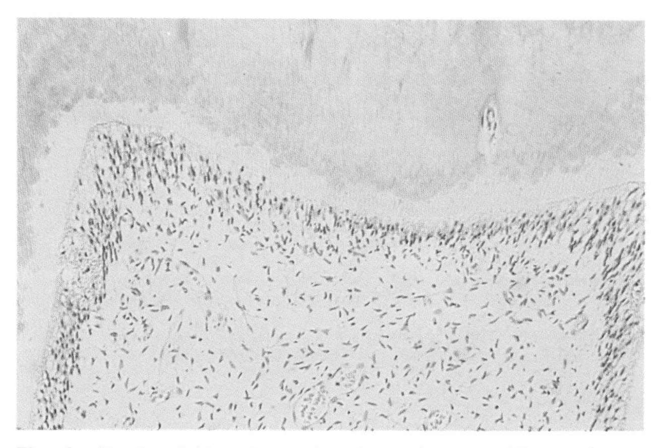

Fig. 1 Dentine bridge 4 months after pulpotomy. The tooth was fractured and had to be extracted as it was not restorable. A normal, uninflamed pulp is shown; new odontoblasts have formed from undifferentiated cells.

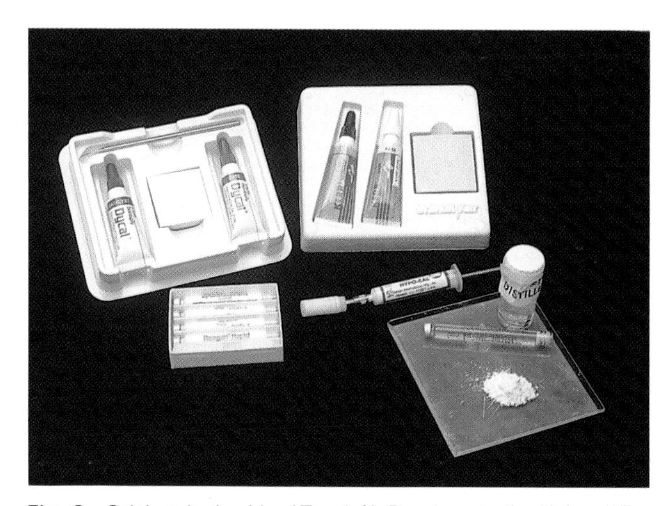

Fig. 2 Calcium hydroxide. (*Top left*) Dycal protective lining. (*Top right*) Life protective lining. (*Bottom left*) Reogan intracanal paste. (*Middle right*) Hypo-cal intracanal paste. (*Bottom left*) $Ca(OH)_2$ powder mixed with distilled water or local anaesthetic solution.

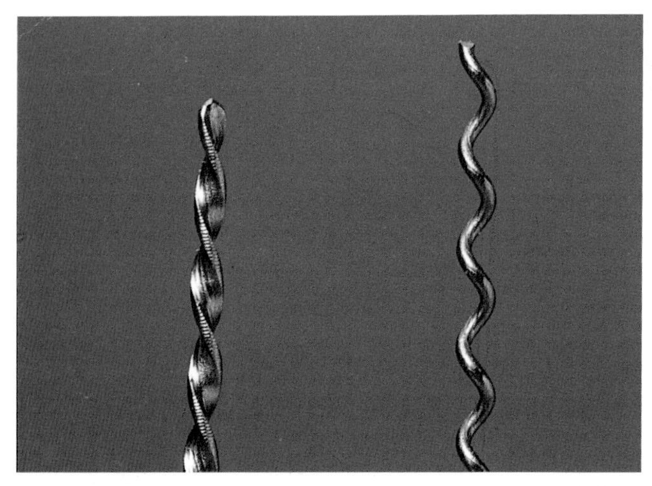

Fig. 3 Spiral root fillers. Blade type of filler (*left*) is less likely to fracture than the conventional round wire filler (*right*).

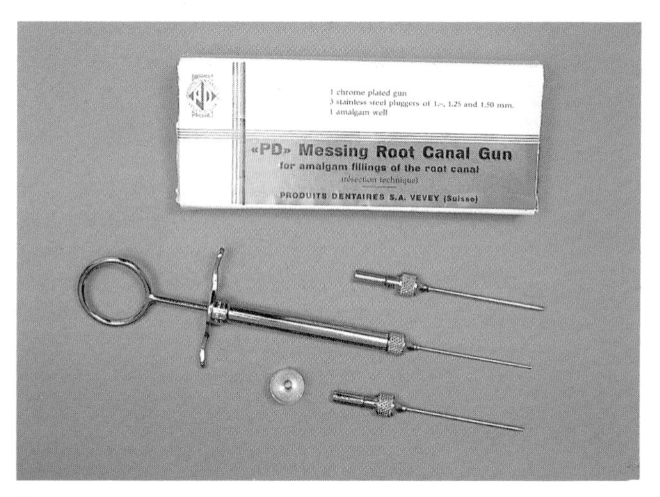

Fig. 4 Messing gun. Using the appropriate sized plugger, calcium hydroxide paste can be inserted to the apical limit.

sary to add a germicide. The advantages of using calcium hydroxide in this form are that variable consistencies may be mixed and a pH of about 12 is achieved, which is higher than that of proprietary brands.

Calcium oxide when mixed with water expands to form calcium hydroxide. An expansion rate in excess of 600% which has been claimed for Biocalex, a proprietary brand of calcium oxide, has been disputed.[4] There is no evidence to show that calcium oxide produces better results than calcium hydroxide.

Root canal sealers containing calcium hydroxide have been introduced (eg Sealapex), the early results of which are promising.

Clinical uses and techniques

The clinical situations where calcium hydroxide may be used in endodontics are discussed below and the techniques described. The method of application of calcium hydroxide to tissue is important if the maximum benefit is to be gained. This applies to pulp capping, pulpotomy, or to an open apex in a pulpless tooth. The exposed tissue should be cleaned thoroughly, any haemorrhage arrested, and the calcium hydroxide placed gently directly on to the tissue, with no debris or blood intervening.[5] A calcium hydroxide cement is applied to protect the pulp in a deep cavity. There are two methods of inserting calcium hydroxide paste into the root canal, the object being to fill the root canal completely with calcium hydroxide so that it is in contact with healthy tissue. Care should be taken to prevent the extrusion of paste into the periapical tissues, although if this does occur healing will not be seriously affected.

Proprietary brands

The canal is first prepared and then dried. A spiral root canal filler, which should be a loose fit in the canal, is selected and the working length of the canal marked on the shank with either marking paste or a rubber stop. The author prefers the blade type of filler (fig. 3), as these are less likely to fracture. The paste of choice is spread evenly on the shank and the spiral filler inserted into the canal and wiped around the walls to reduce air bubble formation. Using a standard handpiece with low rpm, the root canal is filled with paste. Several applications may be required. A pledget of cotton wool is pressed into the pulp chamber so that the paste is condensed further and the access cavity sealed.

Powder–liquid

The powder and liquid are mixed on a glass slab with a spatula to form a thick paste. Although sterile water may be used, local anaesthetic solution is more readily available in the surgery. A standard straight or right-angle amalgam carrier is used to carry the paste to the tooth, and the paste is then ejected into the canal opening. A root canal plugger, which has been preselected to fit the canal but does not wedge against the walls, is used to condense the paste to the full working length. A carrier of smaller diameter, such as a Messing gun, makes the operation simpler by transporting the paste directly to the apical limit of the canal (fig. 4). In narrow canals, a more efficient method is to use a spiral root canal filler. A firmer paste may be made by adding powder to a proprietary brand of calcium hydroxide paste.

Of utmost importance in endodontics is the temporary coronal seal which prevents leakage and (re)contamination of the canal system. Intermediate restorative material (IRM), or amalgam are useful for periods of over 7–10 days; for shorter periods, zinc oxide or Cavit may be used.

Indirect pulp capping

Indirect pulp capping may be used in teeth where carious dentine is lying close to a vital pulp. The aim is to remove infected carious dentine but leave a softened sterile layer over the intact vital pulp. The lining of choice in these cases is a calcium hydroxide cement which would provide a bactericidal effect on any remaining bacteria and encourage the formation of secondary dentine and of a dentine bridge. It is no longer considered necessary to reopen the cavity 2–6 months later and remove the residual softened dentine.[6]

Direct pulp capping

The aim of direct pulp capping is to protect the vital pulp

which has been exposed during cavity preparation or recent trauma. The pulp should be symptom-free and uninfected, and the exposure should be less than 1·0 mm in diameter. Before pulp capping, the vitality of the tooth should be tested with an electronic pulp tester and a radiograph taken to ensure that there is no evidence of pulpal or periapical pathology.

Routine canal medication

Calcium hydroxide paste is not considered necessary as an inter-appointment dressing in routine cases unless there is a special problem or the time between appointments is lengthy. Heithersay[7] recommends that when a chronic periapical lesion is present, a calcium hydroxide dressing should not be placed at the first visit, owing to the possibility of an acute exacerbation.

Long-term dressing

There are occasions when it is not possible to complete a root treatment if the prognosis is questionable, such as in the presence of a large periapical area or because the patient is unable to attend. A calcium hydroxide paste may be used for its bactericidal effect on any organisms which remain in the root canal. In such cases, special care must be taken to provide a good coronal seal.

'Weeping' canal

Sometimes, cases may present with a persistent fluid exudate despite thorough root canal preparation. Placing calcium hydroxide paste in the root canal for a minimum of one week will, in most cases, produce a dry canal ready for obturation.[7]

Root end induction (apexification)

This represents the most widely recognised use of calcium hydroxide in endodontics, and certainly the most dramatic (fig. 5). The cases in which partial or total closure of an open apex can be achieved are:

(1) Vital radicular pulp in an immature tooth pulpotomy (*see* Chapter 7).
(2) Pulpless immature tooth with or without a periapical radiolucent area.
(3) Tooth in which the apex or wall of the root canal has been perforated by an instrument or post.

The success of closure is not related to the age of the patient. It is not possible to determine whether there would be continued root growth to form a normal root apex or merely the formation of a calcific barrier across the apical end of the root. The mode of healing would probably be related to the severity and duration of the periapical inflammation and the consequent survival of elements of Hertwig's sheath.

Inducing apical closure may take anything from 3 to 18 months or occasionally longer. It is necessary to change the calcium hydroxide during treatment; the suggested procedure is given below:

First visit. Thoroughly clean and prepare the root canal. Fill with calcium hydroxide.

Second visit. Two to four weeks later, remove the calcium hydroxide dressing with hand instruments and copious irrigation. Care is taken not to disturb the periapical tissue. The root canal is dried and refilled with calcium hydroxide.

Third visit. Three months later, take a periapical radiograph and root fill if closure is complete. This may be checked by removing the calcium hydroxide and tapping with a paper point against the barrier. Repeat calcium hydroxide dressing if necessary.

Fourth visit. Three months later take a periapical radiograph and root fill if closure is complete. If the barrier is still incomplete repeat calcium hydroxide dressing.

Fifth visit. This should take place 3 to 6 months later. The majority of root closures will have been completed by this time.

Resorption

Calcium hydroxide treatment is recommended in some cases of resorption (See section on resorption.)

Horizontal fractures

Horizontal fractures of the root may be treated, provided the fracture lies within the alveolar bone and does not communicate with the oral cavity. The blood supply may have been interrupted at the fracture site only, so that the apical fragment remains vital. In these cases, the coronal portion of the root can be treated as an open apex. Cvek[8] states that healing with a calcific barrier can be achieved using calcium hydroxide (fig. 6).

Iatrogenic perforations

Iatrogenic perforations are caused by an instrument breaching the apex or wall of the root canal; probably the most common occurrence is during the preparation of a post space. Partial or complete closure by hard tissue may be induced with calcium hydroxide, provided the perforation is not too large, lies within the crestal bone and does not communicate with the oral cavity. Treatment should begin as soon as possible, adopting the same procedure as for root-end induction (fig. 7). Closure of perforations using calcium hydroxide takes considerably longer than root-end induction in most cases. If foreign bodies in the form of root-filling materials, cements or separated instruments have been extruded into the tissues, healing with calcium hydroxide is unlikely to occur and a surgical approach is recommended.

Root resorption

Several different types of resorption are recognised: some are isolated to one tooth and slow spreading, others are rapid, aggressive, and may involve several teeth. Resorption is initiated either from within the pulp, giving rise to internal resorption, or from outside the tooth, where it is termed external resorption.

The aetiology of resorption has been described by Tronstad,[9] who also presented a new classification. In Tronstad's view, the permanent teeth are not normally resorbed, the mineralised tissues are protected by predentine and odontoblasts in the root canal and by precemen-

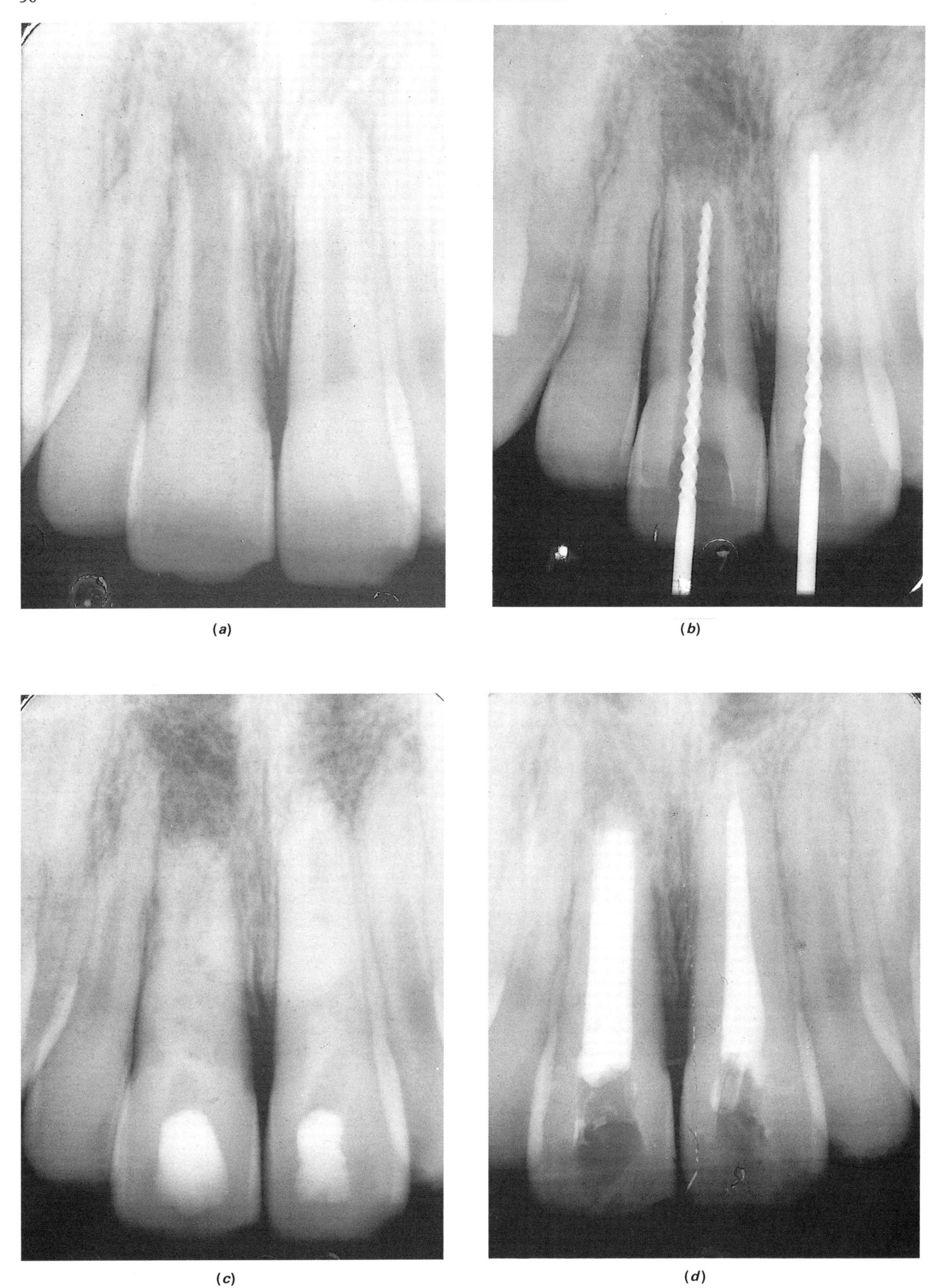

Fig. 5 Root-end induction. (*a*) Pre-operative radiograph taken 8 months following trauma. (*b*) Both pulps were necrotic, the canals were prepared and calcium hydroxide inserted. (*c*) Calcium hydroxide changed after 3 weeks and again 2 months later. Radiograph taken 11 weeks after treatment began. (*d*) Both root fillings placed 2·5 months later.

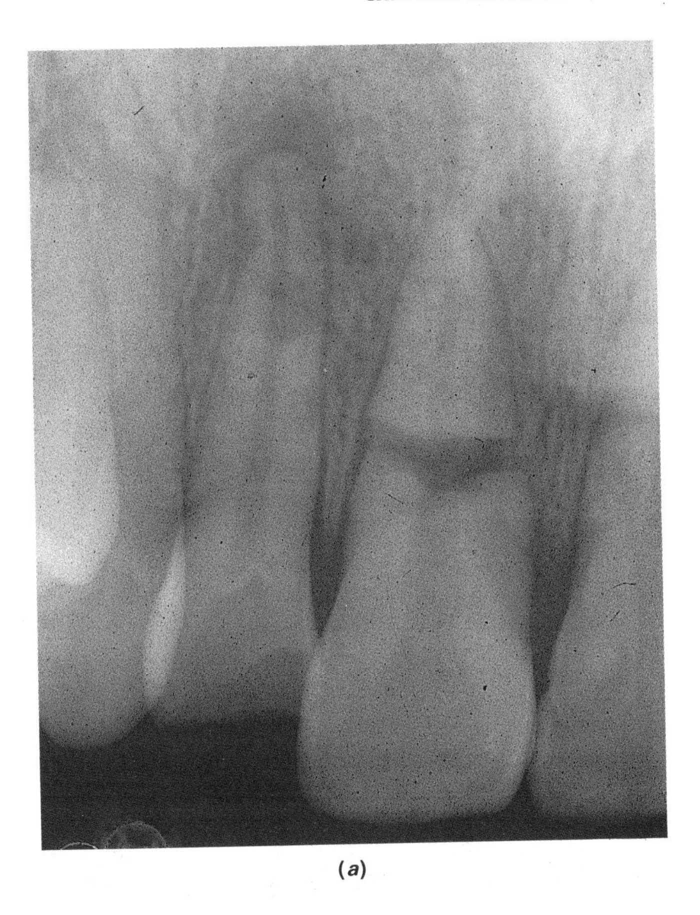
(*a*)

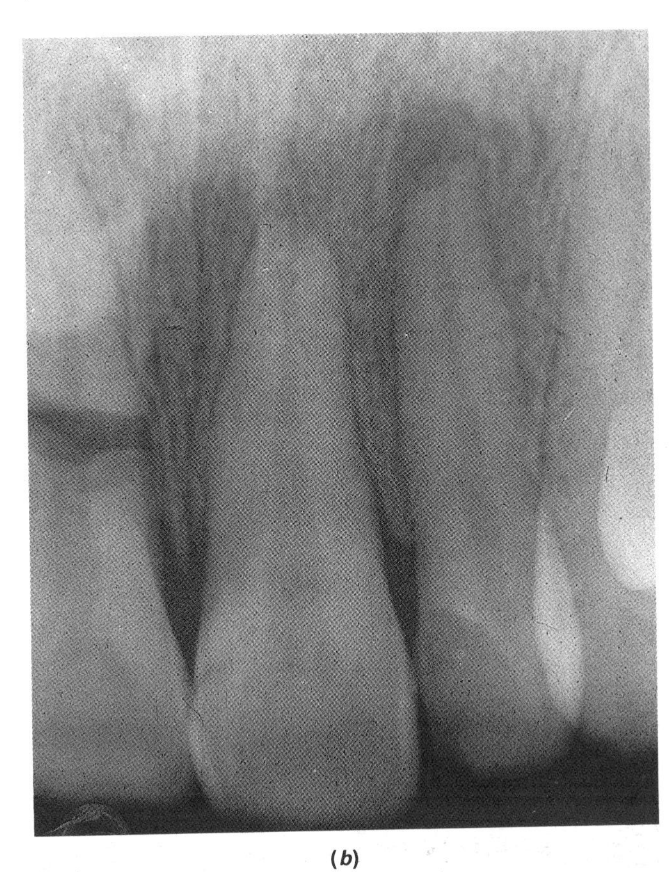
(*b*)

tum and cementoblasts on the root surface. If the predentine or precementum becomes mineralised, or, in the case of the precementum, is mechanically damaged or scraped off, multinucleated cells (fig. 8) colonise the mineralised or denuded surfaces and resorption ensues. Tronstad refers to this type of resorption as inflammatory, which may be transient or progressive. Transient inflammatory resorption will repair with the formation of a cementum-like tissue, unless there is continuous stimulation. Transient root resorption will occur in traumatised teeth or teeth that have undergone periodontal treatment or orthodontics. Progressive resorption may occur in the presence of infection, certain systemic diseases, mechanical irritation of tissue or increased pressure in tissue.

Internal resorption

The aetiology of internal resorption is thought to be the result of a chronic pulpitis. Tronstad[9] believes that there must be a presence of necrotic tissue in order for internal resorption to become progressive. In most cases, the condition is pain-free and so tends to be diagnosed during routine radiographic examination. Chronic pulpitis may follow trauma, caries or iatrogenic procedures such as tooth preparation, or the cause may be unknown. Internal resorption occurs infrequently, but may appear in any tooth; the tooth may be restored or caries-free. The defect may be located anywhere within the root canal system. When it occurs within the pulp chamber, it has been referred to as 'pink spot' because the enlarged pulp is visible through the crown. The typical radiographic appearance is of a smooth and rounded widening of the walls of the root canal.

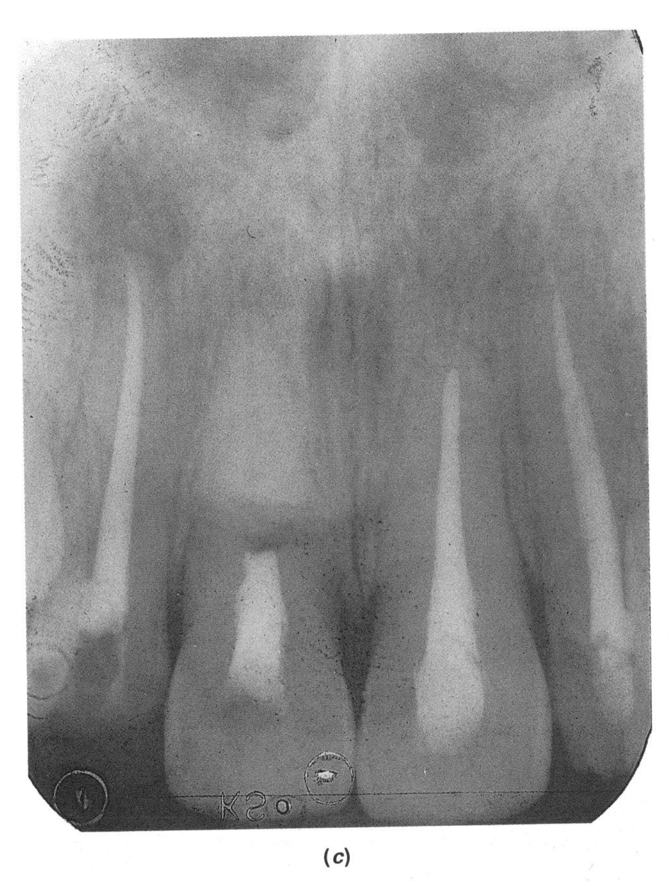
(*c*)

Fig. 6 This female patient aged 23 years was involved in a car accident. Owing to her other injuries, the teeth were not treated for 3 months. (*a*) and (*b*) show both laterals have periapical areas and the 1| (11) is horizontally fractured. The 1| (11) coronal fragment and the |1 (21) were non-vital. All four teeth received calcium hydroxide treatment, the 1| (11) finally being filled 9 months from the first visit. Note that inflammatory resorption in both laterals has resolved.

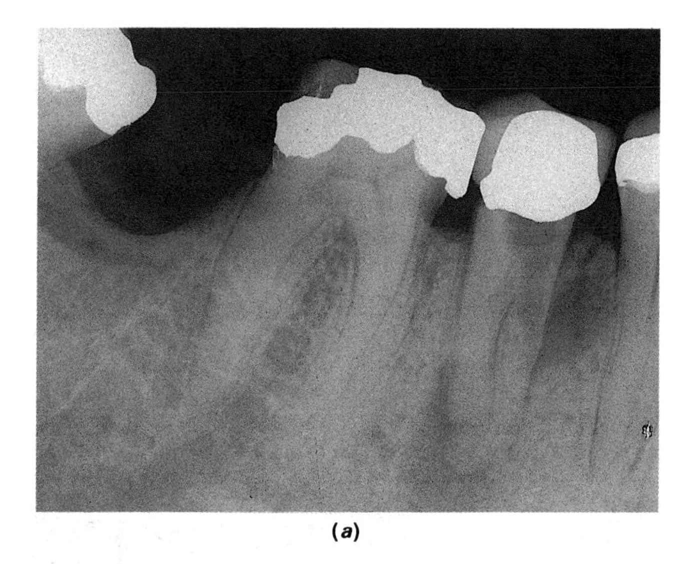

(*a*)

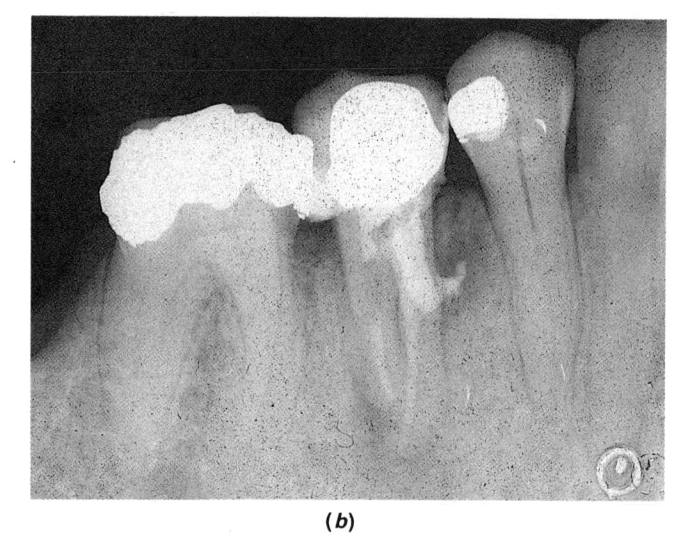

(*b*)

Fig. 7 (*a*) The root of the $\overline{5}|$(45) was perforated while attempting to carry out root canal treatment. (*b*) The calcium hydroxide regime outlined was continued for 12 months and the tooth was finally root-filled. The size of the perforation had reduced but had not healed.

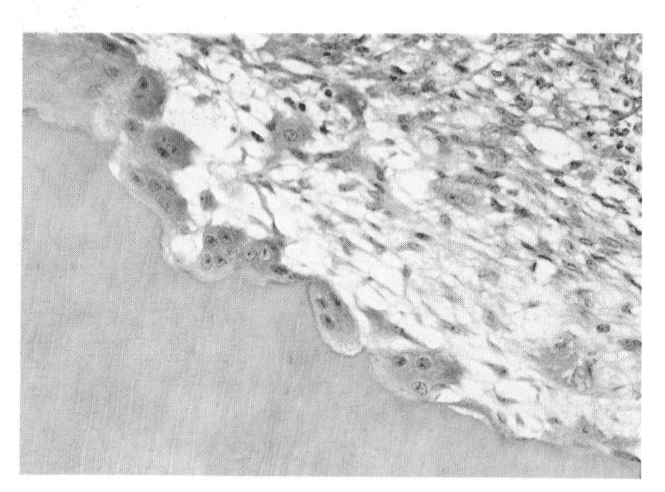

Fig. 8 Resorption lacuna showing multinucleated cells.

If untreated, the lesion is progressive and will eventually perforate the wall of the root, when the pulp will become non-vital (fig. 9). The destruction of dentine may be so severe that the tooth fractures.

The treatment for non-perforated internal resorption is to extirpate the pulp and prepare and obturate the root canal. An interappointment dressing of calcium hydroxide may be used and a warm gutta-percha filling technique helps to obturate the defect (fig. 10). The main problem is the removal of the entire pulpal contents from the area of resorption while keeping the access to a minimum. Hand instrumentation using copious amounts of sodium hypochlorite is recommended. The new ultrasonic technique of root canal preparation may provide a cleaner canal as the acoustic streaming effect removes canal debris from areas inaccessible to the file. The prognosis for these teeth is good and the resorption should not recur.

The treatment of internal resorption that has perforated is more difficult, as the defect must be sealed. When the perforation is inaccessible to a surgical approach, an intracanal seal may be achieved with a warm gutta-percha technique. Before the final root filling is placed, a calcium hydroxide dressing is recommended in one of three ways.

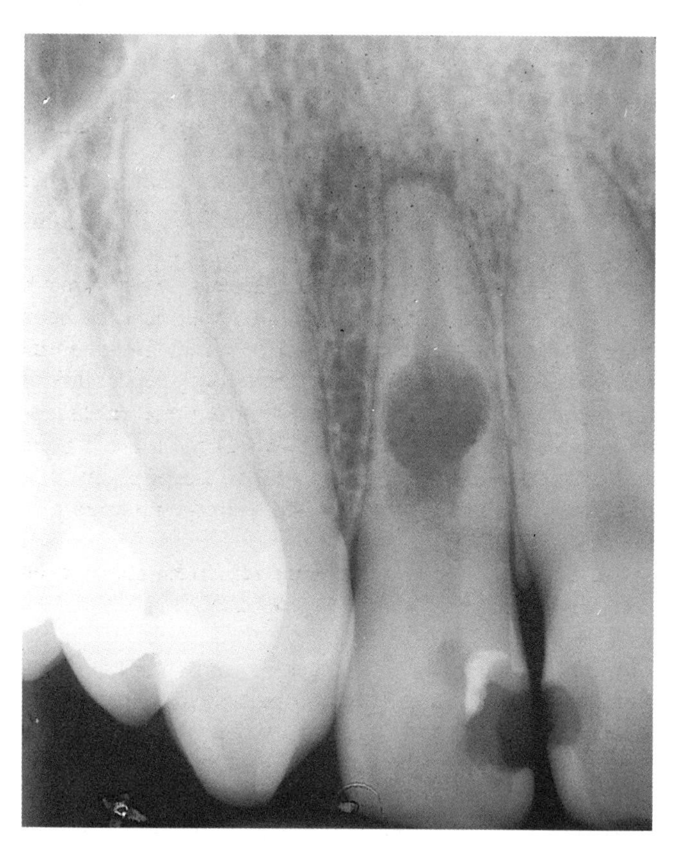

Fig. 9 Typical radiographic view of internal resorption, showing a smooth rounded widening of the root canal.

(1) If there is no associated bone loss, a dry field may be obtained in the root canal by inserting calcium hydroxide for 2–4 weeks.
(2) If the perforation has an associated bone loss, several changes of calcium hydroxide will be necessary over a period of months until bony healing has occurred (fig. 11).
(3) Biological closure of the perforation will usually occur if the calcium hydroxide dressings remain in the root canal for a minimum of 12 months.

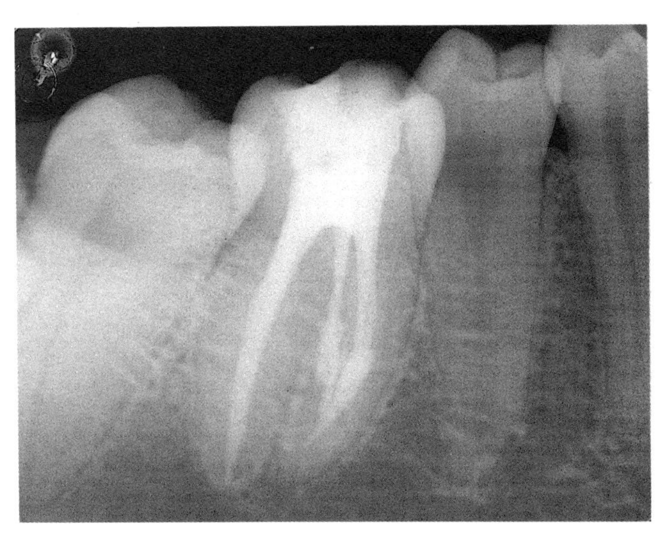

Fig. 10 Internal resorption in both the mesial canals of $\overline{6}|$(46) which has been filled using a warm gutta-percha technique.

External resorption

There are many causes of external resorption, both general and local.[11] An alteration of the delicate balance between osteoblastic and osteoclastic action in the periodontal ligament will produce either a build up of cementum on the root surface (hypercementosis) or its removal together with dentine, which is external resorption.

Resorption may be preceded by an increase in blood supply to an area adjacent to the root surface. The inflammatory process may be due to infection or tissue damage in the periodontal ligament, or alternatively post-traumatic hyperplastic gingivitis and cases of epulis. It has been suggested that osteoclasts are derived from blood-borne monocytes.[12,13] Inflammation increases the permeability of the associated capillary vessels, allowing the release of monocytes which then migrate towards the injured bone and/or root surface. Other causes of resorption include pressure, chemical, systemic diseases and endocrine disturbances.

Six different types of external resorption have been recognised and recorded in the literature.

Surface resorption

Surface resorption is a common pathological finding.[14,15] The condition is self-limiting and undergoes spontaneous repair. The root surface shows both superficial resorption lacunae and repair with new cementum. The osteoclastic activity is a response to localised injury to the periodontal ligament or cementum. Surface resorption is rarely evident on the radiograph.

Inflammatory resorption

Inflammatory resorption is thought to be caused by infected pulp tissue. The areas affected will be around the main apical foramina and lateral canal openings. The cementum, dentine, and adjacent periodontal tissues are involved, and a radiolucent area is visible radiographically (figs 6 and 12).

Further resorption can be prevented by preparing and filling the root canal. If the apex is open, the root canal is first treated with calcium hydroxide.

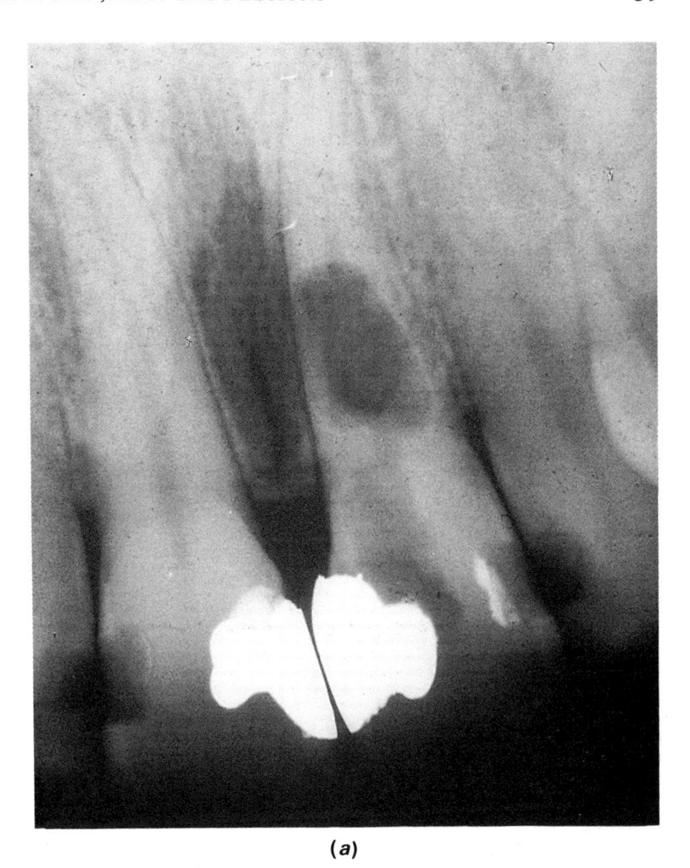

(*a*)

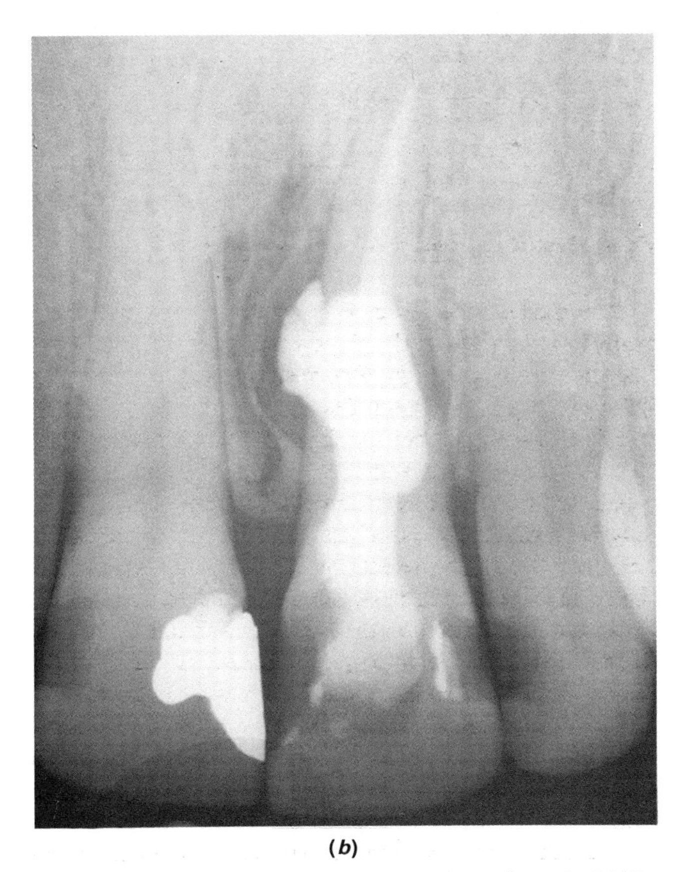

(*b*)

Fig. 11 Internal resorption in the pre-operative radiograph. (*a*) The radiolucent area between the central incisors suggests the resorption has perforated mesially and the pulp is necrotic. The tooth was treated with calcium hydroxide for 6 months before filling with a warm gutta-percha technique. (*b*) The radiolucent area had reduced in size, but the large perforation was still present.

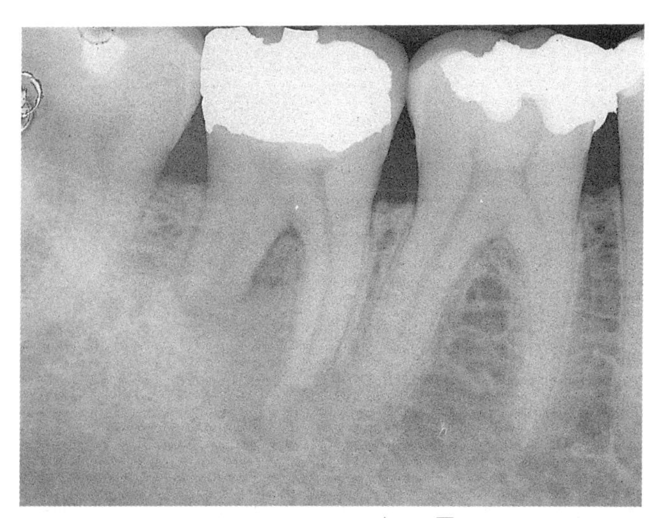

Fig. 12 External inflammatory resorption of $\overline{7}|$(47).

Replacement resorption

Replacement resorption is a direct result of trauma and has been described in detail by Andreasen.[16] A high incidence of replacement resorption follows replantation and luxation, particularly if there was delay in replacing the tooth or there was an accompanying fracture of the alveolus. The condition has also been referred to as ankylosis, because there is gradual resorption of the root, accompanied by the simultaneous replacement by bony trabeculae (fig. 13). Radiographically, the periodontal ligament space is absent, the bone merging imperceptibly with the dentine.

Once started, this condition is usually irreversible, leading ultimately to the replacement of the entire root (fig. 14). Calcium hydroxide treatment is unlikely to help in the treatment of this type of resorption.

Pressure

Pressure on a tooth can eventually cause resorption provided there is a layer of connective tissue between the two surfaces (fig. 15). Pressure can be caused by erupting or impacted teeth, orthodontic movement, trauma from occlusion, or pathological tissue such as a cyst or neoplasm. Resorption due to orthodontic treatment is relatively common. One report[17] in a 5–10 year follow-up after completion of treatment gave an incidence of 28·8% of affected incisors.

It may be assumed that the pressure exerted evokes a release of monocyte cells and the subsequent formation of osteoclasts. If the cause of the pressure is removed, the resorption will be arrested.

Systemic resorption

This may occur in a number of systemic diseases and endocrine disturbances: hyperparathyroidism, Paget's disease, calcinosis, Gaucher's disease and Turner's syndrome. In addition, resorption may occur in patients following radiation therapy.

Idiopathic resorption

There are many reports of cases in which, despite investigation, no possible local or general cause has been found.

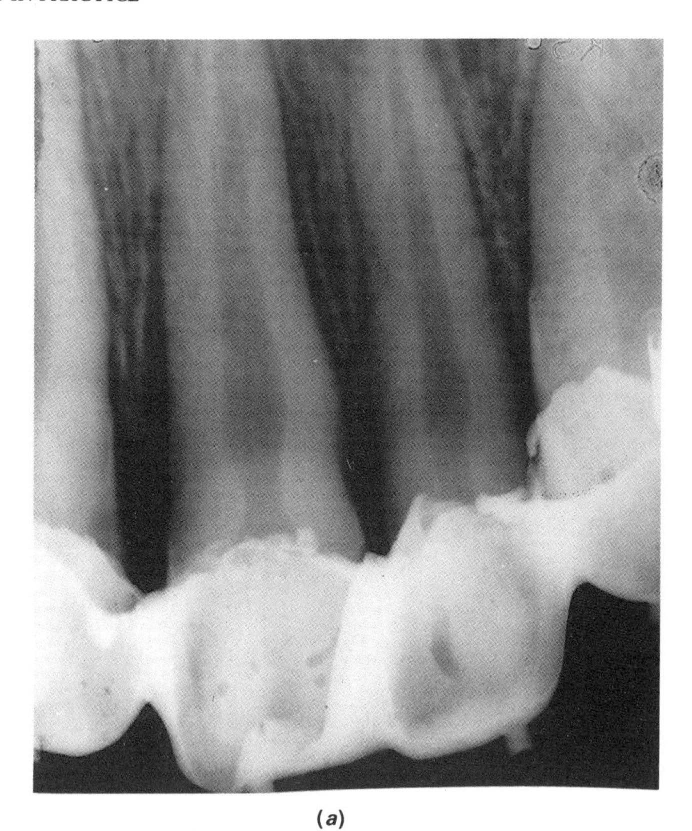

(*a*)

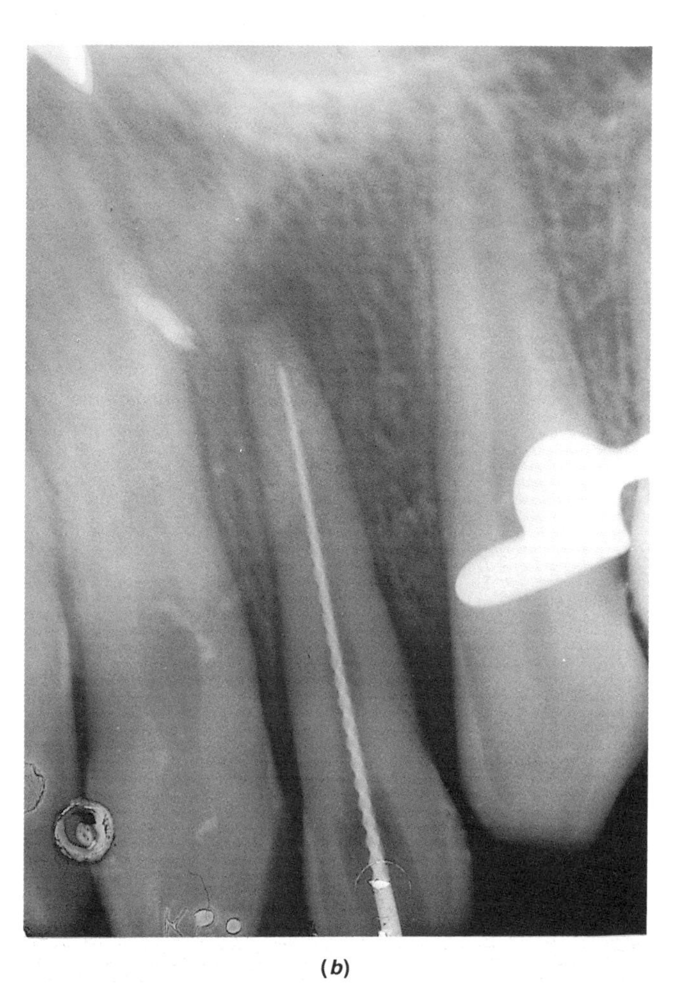

(*b*)

Fig. 13 Replacement resorption. (*a*) A 14-year-old boy playing rugby received a blow which avulsed $\underline{|2}$ (22) and devitalised $\underline{|1}$ (21). The $\underline{|2}$ (22) was immediately replanted and, later on the same day, splinted. (*b*) After calcium hydroxide treatment for 3 months, both teeth were root-filled.

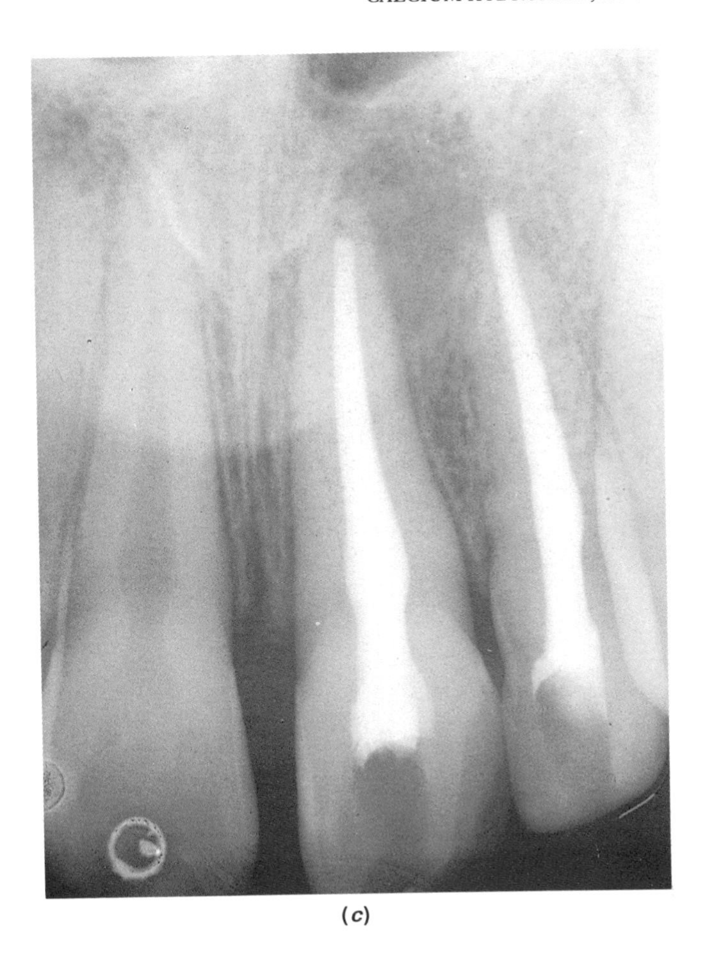

(*c*)

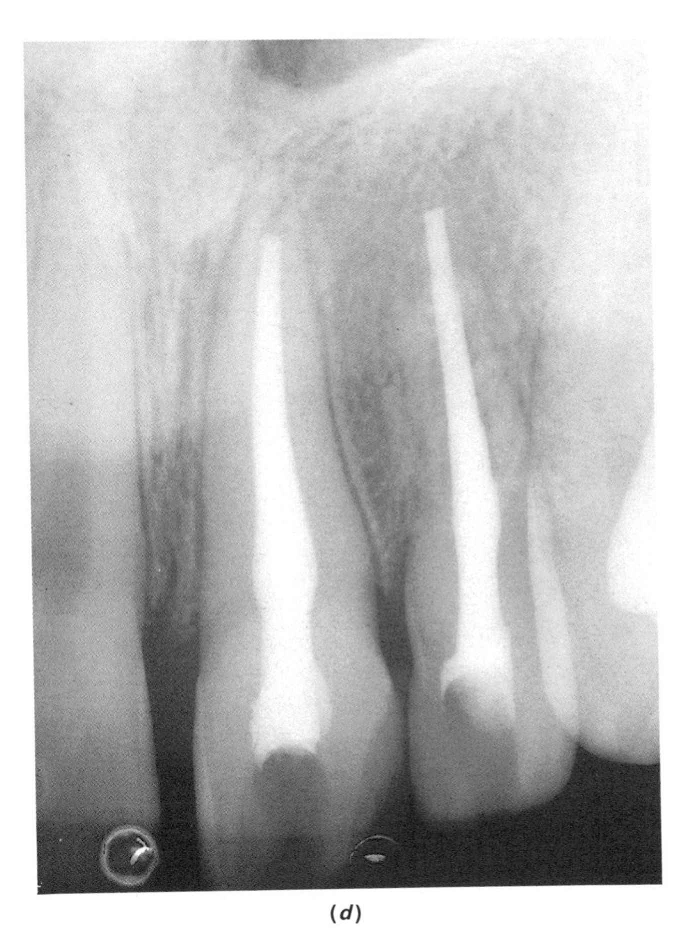

(*d*)

Fig. 13 (*continued*) (*c*) One year later, note the indistinct outline of the |2 (22). (*d*) One year later, the |2 (22) is undergoing replacement resorption; the |1 (21) appears healthy.

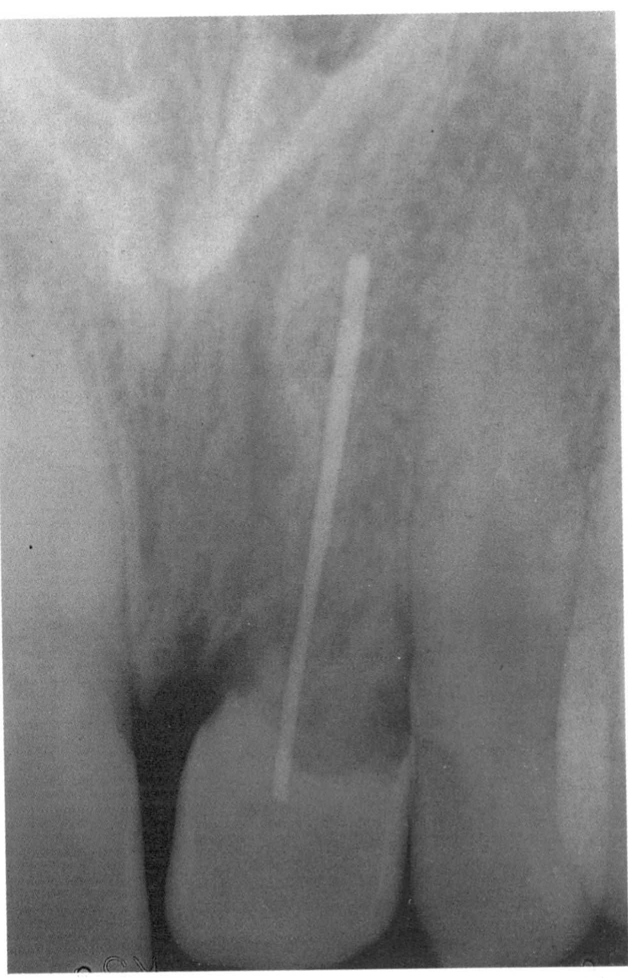

Fig. 14 Replacement resorption, once started, is usually irreversible. This tooth had been avulsed and was root-filled with a gutta-percha point before replanting. The radiograph, taken several years later, shows almost total replacement of the root. The tooth was mobile.

The resorption may be confined to one tooth, or several may be involved. The rate of resorption varies from slow, taking place over years, to quick and aggressive, involving large amounts of tissue destruction over a few months. The site and shape of the resorption defect also varies. Two different types of idiopathic resorption have been described.

Apical resorption is usually slow and may arrest spontaneously; one or several teeth may be affected, with a gradual shortening of the root, while the root apex remains rounded (fig. 16).

Cervical external resorption takes place in the cervical area of the tooth. The defect may form either a wide shallow crater (fig. 17) or, conversely, a burrowing type of resorption (fig. 18). This latter type has been described variously as peripheral cervical resorption, burrowing resorption, pseudo pink spot, resorption extra camerale, and, more recently, extracanal invasive.

There is a small defect on the external surface of the tooth; the resorption then burrows deep into the dentine with extensive tunnel-shaped ramifications. It does not, as a rule, affect the dentine and predentine in the immediate vicinity of the pulp. This type of resorption is easily mistaken for internal resorption. Cervical resorption may

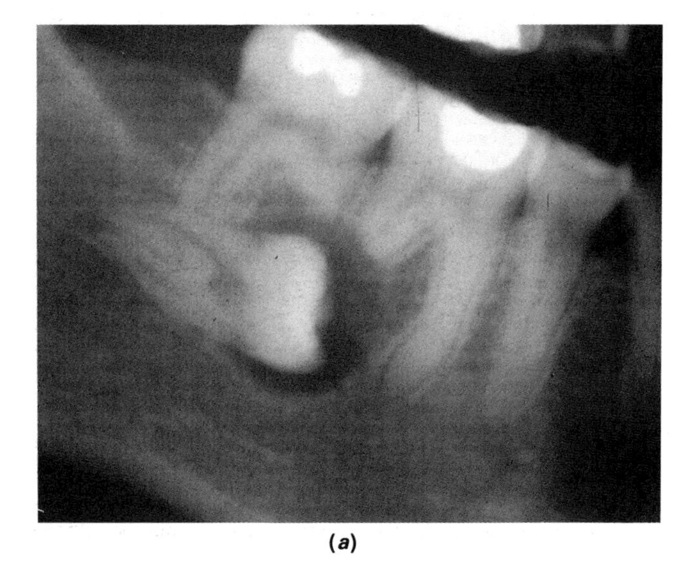

(*a*)

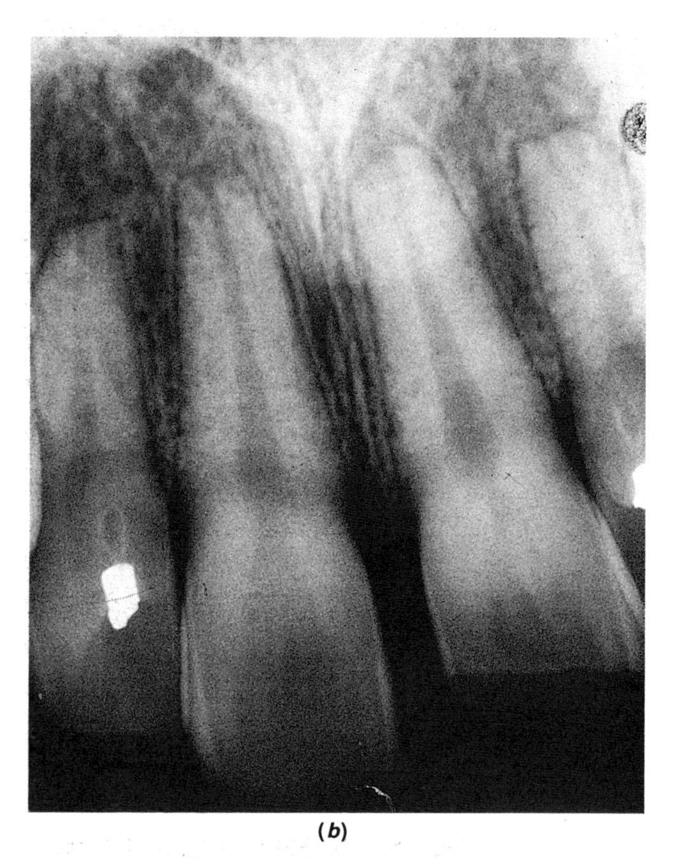

(*b*)

Fig. 15 Pressure resorption. (*a*) Resorption due to an impacted tooth. (*b*) The apices of the teeth have a 'cut off' appearance. This is typical of resorption due to orthodontic treatment.

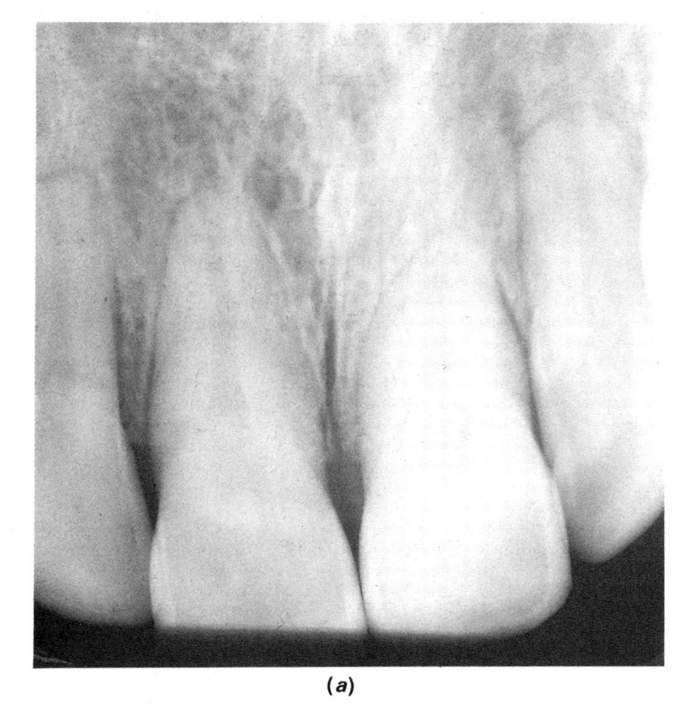

(*a*)

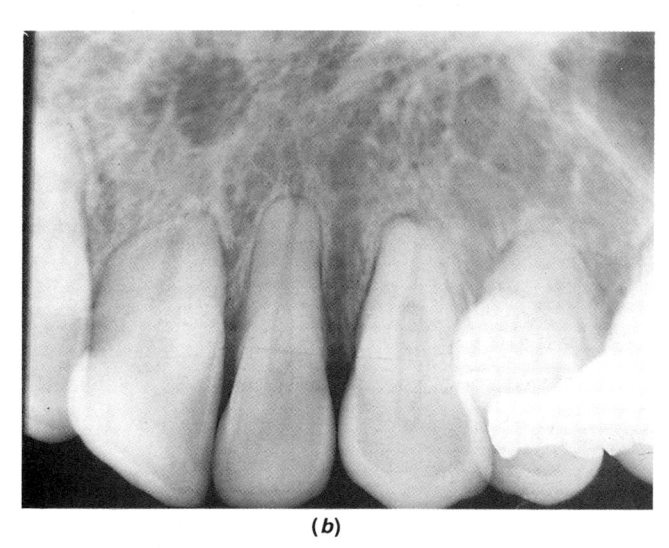

(*b*)

Fig. 16 (*a*) and (*b*). Idiopathic apical resorption. Female patient aged 35 years had generalised shortening of roots. The teeth were vital, although several showed pulpal calcification. Despite extensive investigation, no cause was found.

be caused by chronic inflammation of the periodontal ligament or by trauma. Both types of cervical resorption are best treated by surgical exposure of the resorption lacunae and removal of the granulation tissue. The resorptive defect is then shaped to receive a restoration.

The perio-endo lesion

The differential diagnosis of perio-endo lesions has become increasingly important as the demand for complicated restorative work has grown. Neither periodontic nor endodontic treatment can be considered in isolation as clinically they are closely related and this must influence the diagnosis and treatment. The influence of infected and necrotic pulp on the periapical tissues is well known, but there remains much controversy over the effect that periodontal disease could have on a vital pulp.

Examination of the anatomy of the tooth shows that there are many paths to be taken by bacteria and their toxic products between the pulp and the periodontal ligament. Apart from the main apical foramina, lateral canals exist in approximately 50% of teeth.[18-20] Seltzer[21] induced interradicular periodontal changes in dogs and monkeys by inducing pulpotomies and concluded that noxious material passed through dentinal tubules in the floor of the pulp chamber.

In addition to dentinal tubules, microfractures are often present in teeth, allowing the passage of microorganisms. Clinically, it is common to see cervical sensitivity.

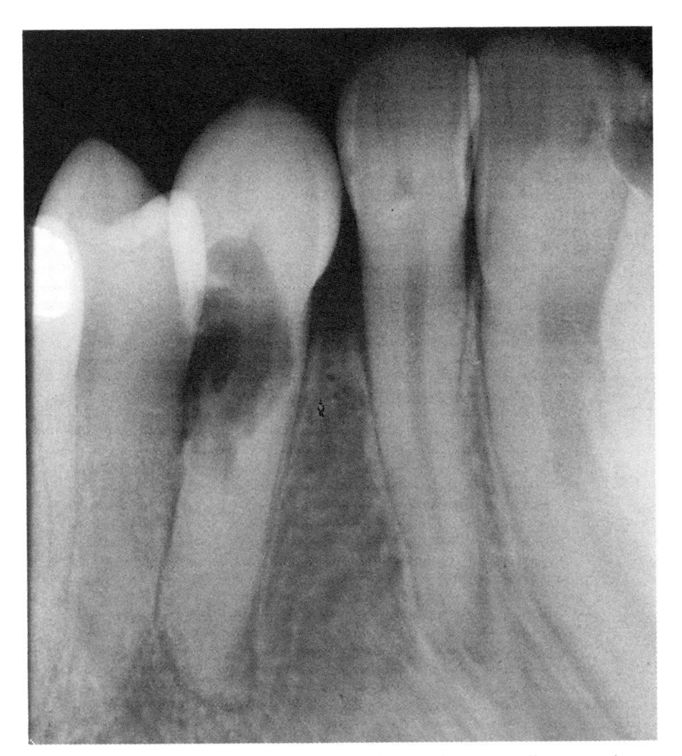

Fig. 17 Cervical external resorption. The defect was discovered at a routine check. The patient had no history of trauma. The tooth was pulpless; extraction was recommended.

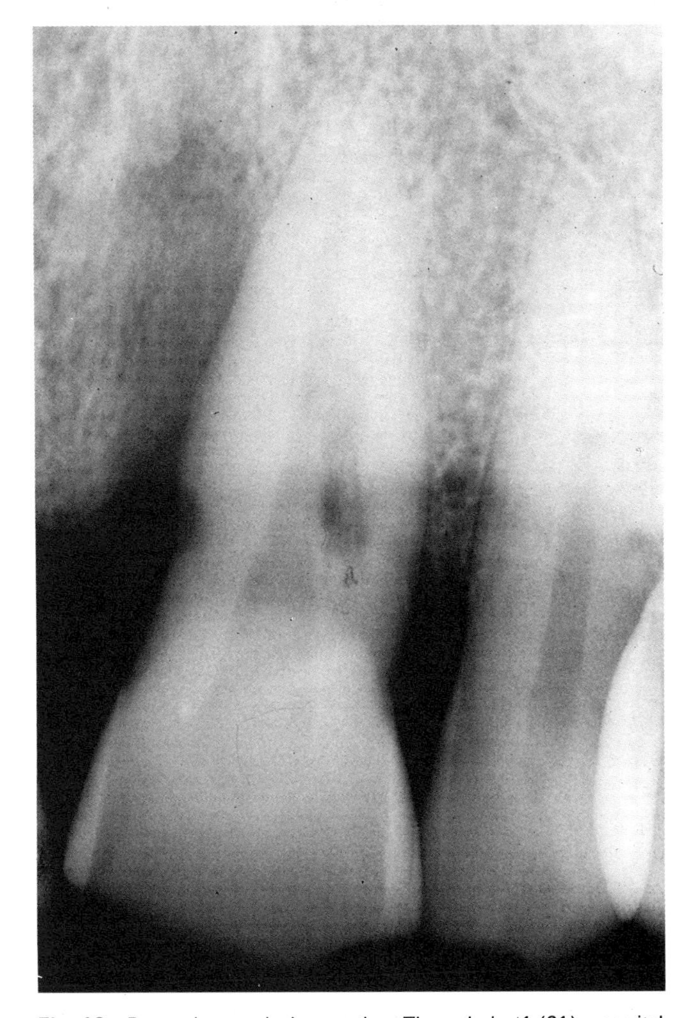

Fig. 18 Burrowing cervical resorption. The pulp in ⌊1 (21) was vital, but the tooth had an isolated 7·0 mm pocket mesially. The pulp canal is visible, suggesting that the defect is external in origin.

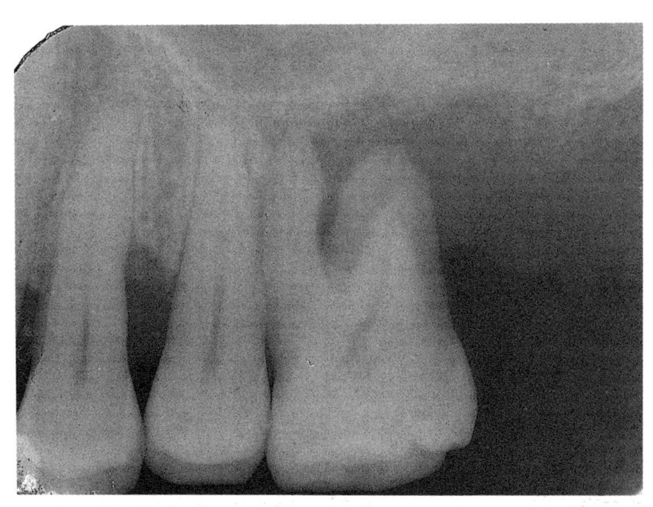

Fig. 19 Extensive perio-endo lesion. Despite extensive bone loss and pocketing to the apex on the distobuccal and palatal roots, the pulp still gave a vital response.

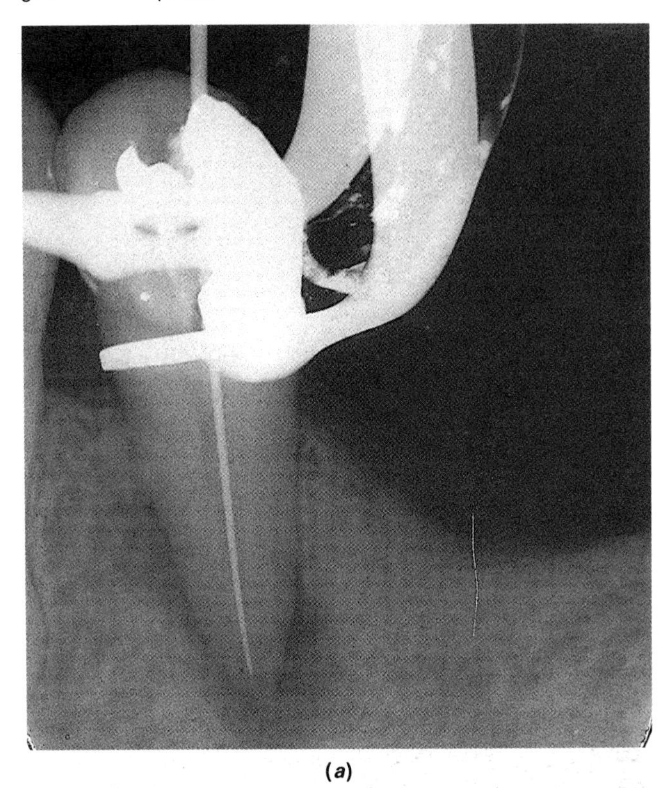

(*a*)

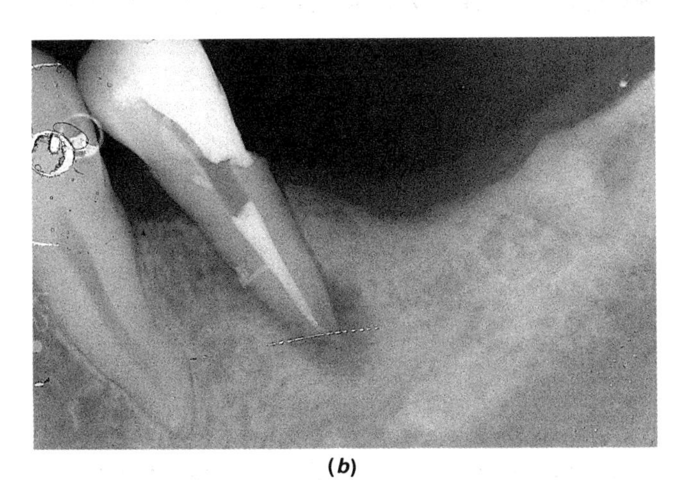

(*b*)

Fig. 20 Class 1 perio-endo lesion. (*a*) The ⌈4 (34) had a necrotic pulp and a pocket mesially. (*b*) A lateral canal is visible on the mesial aspect of the root.

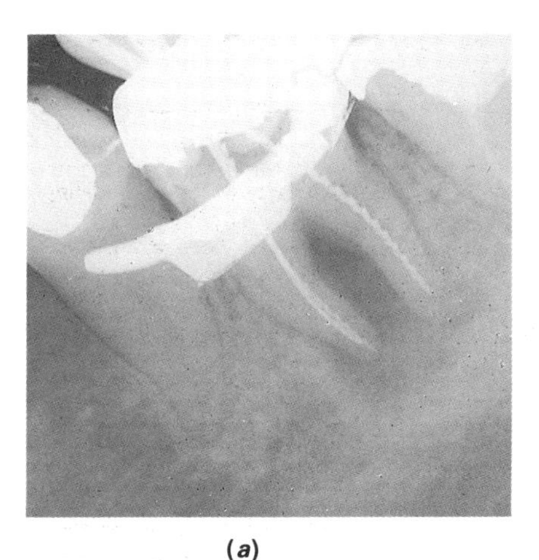

(*a*)

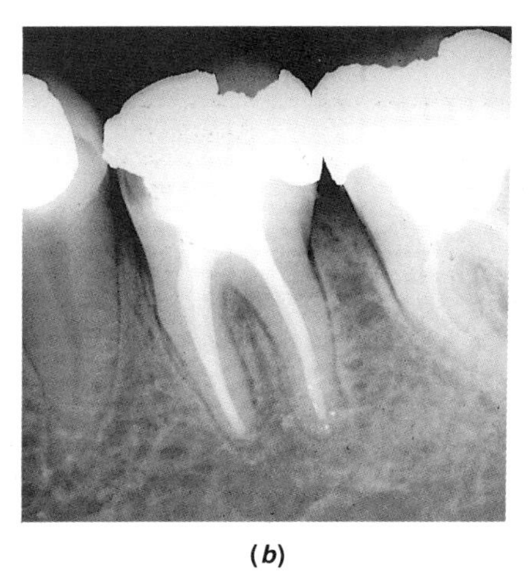

(*b*)

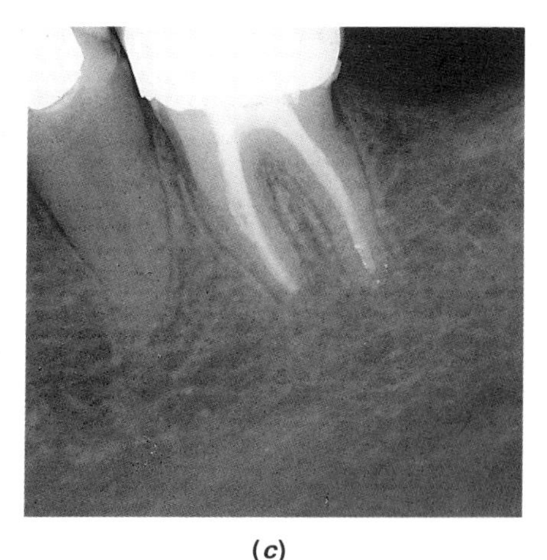

(*c*)

Fig. 21 Class 1 perio-endo lesion. (*a*) The |6 (36) is non-vital with furcation involvement. The tooth was root-filled. (*b*) One year later, bony healing is complete in the furcation. (*c*) Four years later, there has been no breakdown around the |6 (36), but the |7 (37) had been extracted.

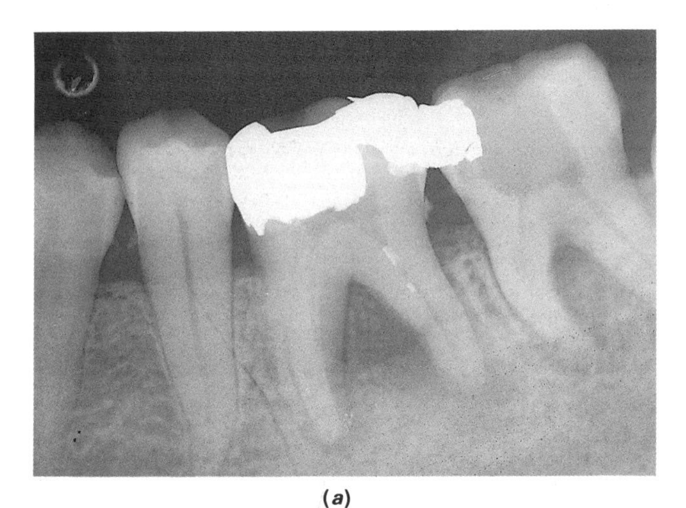

(*a*)

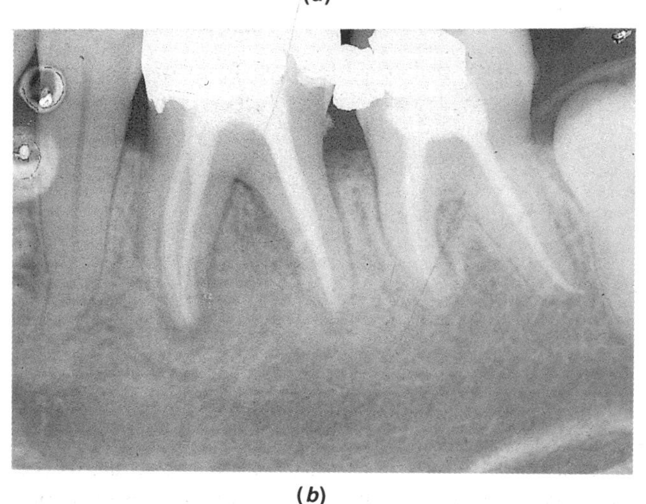

(*b*)

Fig. 22 Class 2 perio-endo lesion. (*a*) The |6 (36) and |7 (37) have furcation involvement. Note also the calculus on the distal aspect of |6 (36). (*b*) Follow-up radiograph taken one year later shows bony healing, but clinically, periodontal disease is still evident, as the patient had not sought further treatment.

The controversy concerning the effect of periodontal disease on the pulp ranges from Rubach and Mitchell,[18] who demonstrated that pulpitis or pulp necrosis or both can occur as a result of periodontal inflammation, to Mazur and Massler[22] who categorically stated that, from the results of their investigation, pulpal changes are independent of the status of the periodontium. In the author's opinion, Langeland[23] presents the most rational view, which is that periodontal disease may damage pulp tissue via accessory or lateral canals, but total pulpal disintegration will not occur unless all the main apical foramina are involved by bacterial plaque (fig. 19).

The problem that faces the clinician treating perio-endo lesions is to assess the extent of the disease and to decide whether the tooth or the periodontium is the primary cause. Only by carrying out a careful examination can the operator judge the prognosis and plan the treatment. There are several ways in which perio-endo lesions can be classified; the one given below is a slight modification of the Simon, Glick and Frank classification.[24]

Classification of perio-endo lesions

Class 1. Primary endodontic lesion draining through the periodontal ligament

Class 1 lesions present as an isolated periodontal pocket or swelling beside the tooth. The patient rarely complains of pain, although there will often be a history of an acute episode.

The cause of the pocket is a necrotic pulp draining through the periodontal ligament (fig. 20). The furcation area of both premolar and molar teeth may be involved. Diagnostically one should suspect a pulpally-induced lesion when the crestal bone levels on both the mesial and distal aspects appear normal and only the furcation shows a radiolucent area (fig. 21) (Table I).

Class 2. Primary endodontic lesion with secondary periodontal involvement

If left untreated, the primary lesion may become secondarily involved with periodontal breakdown. A probe may encounter plaque or calculus in the pocket. The lesion will resolve partially with root canal treatment but complete repair will involve periodontal therapy. A Class 2 lesion is shown in figure 22 (Table II).

Class 3. Primary periodontal lesions

Class 3 lesions are caused by periodontal disease gradually spreading along the root surface. The tooth will remain

Table I Diagnosis and treatment of Class I perio-endo lesions

Diagnosis	Treatment
(1) Necrotic pulp with discharge via periodontal ligament	(1) Root canal treatment
(2) Isolated lesion	(2) Review after 4–6 months for healing of periodontal pocketing and bone repair
(3) Good crestal bone levels on both the mesial and distal aspects of the tooth	(3) Prognosis good
(4) Root surface in pocket contains minimal plaque or calculus	

Table II Diagnosis and treatment of Class 2 perio-endo lesions

Diagnosis	Treatment
(1) Necrotic pulp or other source of irritation to the periodontium from the root such as a poorly obturated root filling or root fracture	(1) Root treatment or re-root treatment
(2) Isolated deep pocket, although there may be generalised periodontal disease	(2) Periodontal treatment will be required
(3) Periodontal breakdown with calculus or plaque present in the pocket	(3) Vertical fractures are invariably untreatable and the tooth is extracted
	(4) Prognosis generally good

vital, although in time there will be some degenerative pulpal changes. The tooth may become mobile as the attachment apparatus and surrounding bone are destroyed, leaving deep periodontal pocketing. There is usually periodontal disease involving other areas in the mouth, except where there is a local predisposing factor such as a defective restoration, poor proximal contact or a developmental groove. Two examples of a Class 3 lesion are given in figures 23 and 24 (Table III).

Class 4. Primary periodontal lesions with secondary endodontic involvement

A Class 3 lesion progresses to a Class 4 lesion with the involvement of the main apical foramina or possibly a large lateral canal. It is sometimes difficult to decide whether the lesion is primarily endodontic with secondary periodontal involvement (Class 2), or primarily periodontal with secondary endodontic involvement (Class 4), particularly in the late stages (fig. 25). If there is any doubt, the necrotic pulp should be removed; any improvement indicates Class 2 (Table IV).

Root removal and root canal treatment

To prevent further destruction of the periodontium in multi-rooted teeth, it may be necessary to remove one or occasionally two roots. As this treatment will involve root canal therapy and periodontal surgery, the operator must consider the more obvious course of treatment, which is to

Table III Diagnosis and treatment of Class 3 perio-endo lesions

Diagnosis	Treatment
(1) Pulp is vital	(1) Oral hygiene advice, scaling and root planing
(2) Description of the pain is periodontal rather than pulpal. It is a chronic localised discomfort, sometimes relieved by biting on the tooth	(2) Removal of cause, such as poor restoration. Developmental grooves, once they are periodontally involved, are difficult to treat successfully
(3) Periodontal breakdown with calculus or plaque present in the pocket	(3) Periodontal surgery may be indicated
	(4) Prognosis becomes worse as the disease advances

Table IV Diagnosis and treatment of Class 4 perio-endo lesions

Diagnosis	Treatment
(1) Non-vital pulp	(1) Root treatment or re-root treatment
(2) Long-standing periodontal involvement with deep pocketing to the root apex. Plaque and calculus present in the pocket	(2) Periodontal treatment will be required in some cases with surgery
(3) Discharge usually present from the pocket on palpation	(3) Prognosis poor
(4) Tooth mobile in the majority of cases	
(5) Generalised periodontal disease often present, except in the case of a palatal groove or root fracture	

extract the tooth and provide some form of fixed prosthesis. As a guide, the following factors should be considered before root resection.

(1) *Functional tooth*. The tooth should be a functional member of the dentition.

(2) *Root filling*. It should be possible to provide root canal treatment which has a good prognosis. In other words the root canals must be fully negotiable.

(3) *Anatomy*. The roots should be separate with some interradicular bone so that the removal of one root will not damage the remaining root(s). Access to the tooth must be sufficient to allow the correct angulation of the handpiece to remove the root. A small mouth may contra-indicate the procedure.

(4) *Restorable*. Sufficient tooth structure must remain to allow the tooth to be restored. The finishing line of the restoration must be envisaged to ensure that it will be cleansable by the patient.

(5) *Patient suitability*. The patient must be a suitable candidate for the lengthy operative procedures and be able to maintain a high standard of oral cleanliness around the sectioned tooth.

A tooth that requires a root to be resected will need root

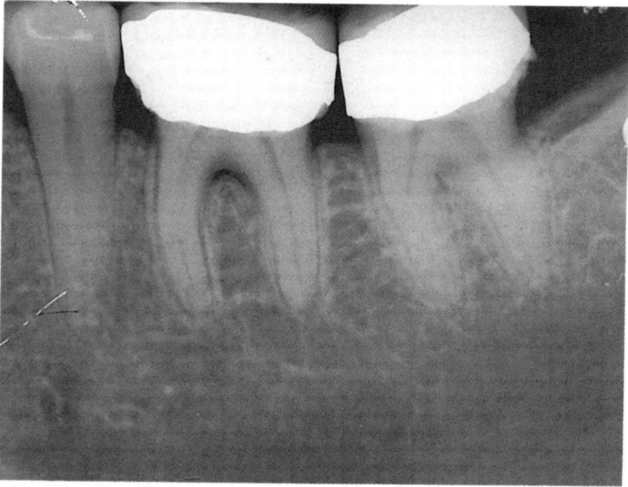

Fig. 23 Class 3 perio-endo lesion. The patient presented with mild discomfort from |6 (36). The radiograph shows an early furcation area. The crown had a large overhang buccally and the tooth was vital. It was recommended that the crown should be replaced.

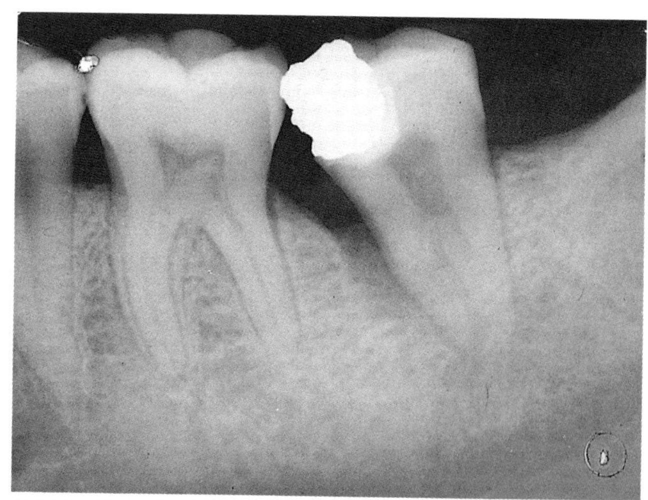

Fig. 24 Class 3 perio-endo lesion. The |7 (37) shows extensive bone loss around the mesial root, with 7·0 mm pocketing. The tooth responded vitally to the pulp tests.

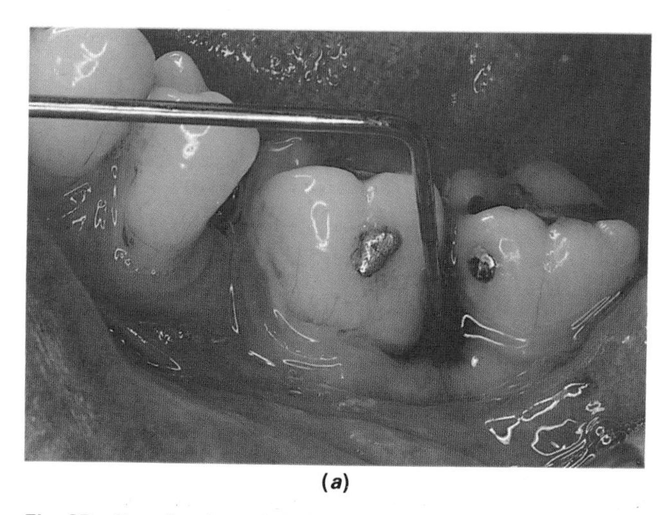

(*a*)

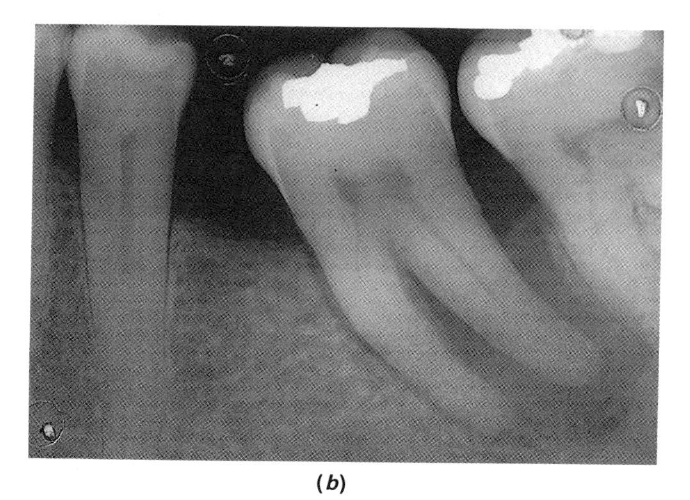

(*b*)

Fig. 25 Class 4 perio-endo lesion. (*a*) and (*b*) There was pocketing to the distal root apex of the |6 (36) and the pulp did not respond to the electric pulp tester. It was decided to extract the tooth and provide a bridge.

canal treatment. The surgery must be planned with care, particularly with respect to the timing of the root treatment. Ideally, the tooth should be root-filled prior to surgery, except for the root which is for resection. The pulp is extirpated from the root which is to be removed and the canal widened in the coronal 2–3 mm and packed with amalgam (fig. 26). This means a retrograde filling will not have to be placed at the time of surgery—a procedure which is difficult to perform owing to poor access and blood contamination of the filling and the likelihood of amalgam falling into the socket. Sometimes it is not possible to decide which root should be removed until the surgical flap has been reflected. The suggested procedure in these cases is as follows.

Vital teeth

Either resect the root and leave the pulp exposed[25] (the root filling should be started 10–14 days later), or extirpate the pulp prior to surgery and place a retrograde dressing of zinc oxide at the time of operation.

Pulpless teeth

The root canals should be prepared and root-filled and a coronal dressing placed. In these cases it is recommended to wait for a period of 2–3 months before a decision on the surgery is made. In some Class 2 lesions there may be sufficient resolution of bone to make surgery unnecessary.

References

1 Hermann B W. Dentinobleration der Wurzelkanale nach der Behandlung mit Kalcium. *Zahnarzt Rundschau* 1930; **39:** 888.
2 Sciaky I, Pisanti S. Localisation of calcium placed over amputated pulps in dog's teeth. *J Dent Res* 1960; **39:** 1128–1132.
3 Tronstad L, Andraeson J O, Hasselgren G, Kristerson L, Riis I. pH changes in dental tissues after root canal filling with calcium hydroxide. *J Endod* 1981; **7:** 17–21.
4 Donnelly L L, Harty F J. An *in-vitro* investigation of 'Biocalex 6–9'. *Endodontics* 1979; **12:** 25–31.
5 Schroder V. Effect of an extra-pulpal blood clot on healing following experimental pulpotomy and capping with calcium hydroxide. *Odont Revy* 1973; **24:** 1–11.
6 Plasschaert A J M. The treatment of vital pulps. 2. Treatment to maintain pulp vitality. *Int Endod J* 1983; **16:** 115–120.
7 Heithersay G S. Calcium hydroxide in the treatment of pulpless teeth with associated pathology. *J Br Endod Soc* 1975; **8:** 74–92.

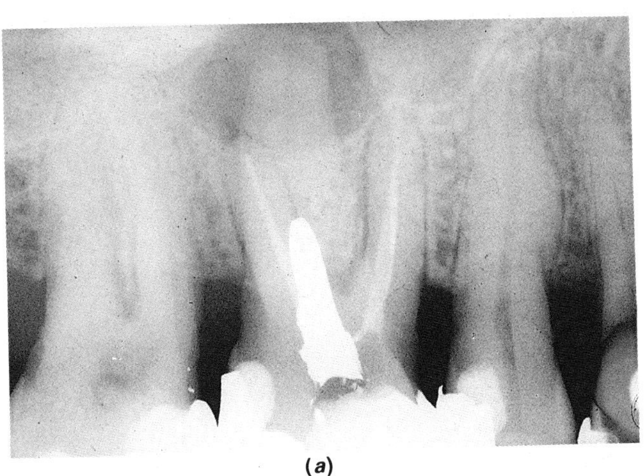

(*a*)

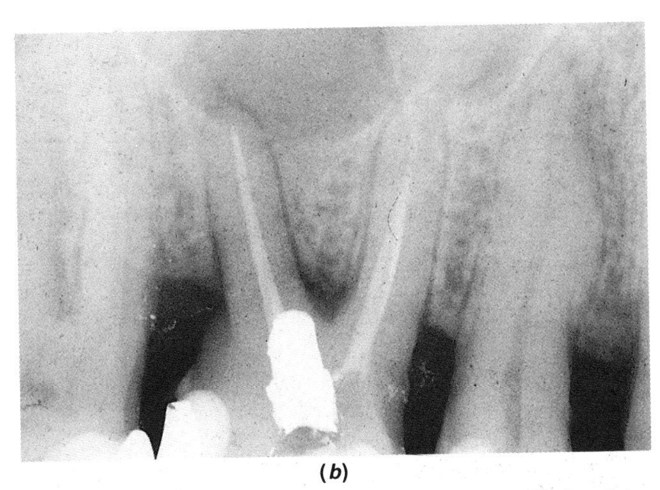

(*b*)

Fig. 26 Root resection. (*a*) The palatal root of the 6|(16) was periodontally involved, with deep pocketing. The buccal roots were filled and the palatal root was prepared to half its length, using Gates-Glidden drills and packed with amalgam. (*b*) A flap was reflected and the palatal root removed.

8 Cvek M. Treatment of non-vital permanent incisors with calcium hydroxide. IV. Periodontal healing and closure of the root canal in the coronal fragment of teeth with intra-alveolar fracture and vital apical fragment. A follow-up. *Odont Revy* 1974; **35:** 239.

9 Tronstad L. Root resorption—etiology, terminology and clinical manifestations. *Endod Dent Traumatol* 1988; **4:** 241–252.

10 Ingle J, Taintor J. *Endodontics*. 3rd ed. p 411. Philadelphia: Lea and Febiger, 1985.

11 Pindborg J J. *Pathology of the dental hard tissues*. pp 338. Copenhagen: Munksgaard, 1970.

12 Newman W G. Possible etiologic factors in external root resorption. *Am J Orthod* 1975; **67:** 522–539.

13 Brown W A B. Resorption of permanent teeth. *Br J Orthod* 1982; **9:** 212–220.

14 Massler M, Malone A J. Root resorption in human permanent teeth. A roentgenographic study. *Am J Orthod* 1954; **40:** 619–633.

15 Henry J L, Weinmann J P. The pattern of resorption and repair of human cementum. *J Am Dent Assoc* 1951; **42:** 270–290.

16 Andraesen J O. *Traumatic injuries of the teeth*. 2nd ed. p 211. Copenhagen: Munksgaard, 1981.

17 Cwyk F, Sciat-Pierre F, Tronstad L. Endodontic implications of orthodontic tooth movement. *J Dent Res* 1984; **63:** (IADR abstract, no. 1039).

18 Rubach W C, Mitchell D F. Periodontal disease, accessory canals and pulp pathosis. *J Periodontol* 1965; **34:** 38.

19 Lowman J V, Burke R S, Pellen G B. Patent accessory canals: Incidence in molar furcation region. *Oral Surg* 1973; **36:** 580–584.

20 Burch J G, Hulen S. A study of the presence of accessory foramina and the topography of molar furcations. *Oral Surg* 1974; **38:** 451–455.

21 Seltzer S, Bender I B, Nazimor M, Sinai I. Pulpitis-induced interradicular periodontal changes in experimental animals. *J Periodontol* 1967; **38:** 124.

22 Mazur B, Massler M. Influence of periodontal disease on the dental pulp. *Oral Surg* 1964; **17:** 592–603.

23 Langeland K, Rodrigues H, Dowden W. Periodontal disease, bacteria, and pulpal histopathology. *Oral Surg* 1974; **37:** 252–270.

24 Simon J H S, Glick D H, Frank A L. The relationships of endodontic–periodontic lesions. *J Periodontol* 1972; **43:** 202–208.

25 Tagger M, Smukler N. Microscopic study of the pulps of human teeth following vital root resection. *Oral Surg* 1977; **44:** 96–105.

Biocalex: Laboratoire Spad, 21800 Quetigsy, BP7 Dijon, France.
Dycal: Caulk Dentsply, PO Box 359, Milford, Delaware, 19963, USA.
Hawes-Neos root filler: 6925 Gentillino, Lugano 3, Switzerland.
Hypo-cal: Ellman International, Hewlett, New York 11557, USA.
Life Protective Lining: Kerrs Manufacturing Company, Detroit 8, Michigan, USA.
Messing Gun: Produits Dentaires SA, Vevey, Switzerland.
Reogan: Vivadent, Schaan, Liechtenstein.
Sealapex: Kerrs Divisions of Sybron Corporation, 28200 Wick Road, Box 445, Romulus, Michigan 48174, USA.

9

Endodontic Treatment for Children

The considerations and difficulties to be taken into account when contemplating endodontic treatment in children are, at times, different from those encountered in adults.

The aims of therapy include the removal of infection and chronic inflammation and thus the relief of associated pain. The maintenance of arch length is important for good masticatory function and the future eruption of the permanent dentition with optimal development of the occlusion. As a general rule it is preferable to conserve a tooth rather than carry out an extraction, but if this becomes necessary, balanced extractions should always be kept in mind. However, there are important assessments to be made as to the patient's suitability for treatment. The general health of the patient should be checked to ensure that there are no contra-indications to endodontic therapy. The attitude of the parent to treatment and the child's ability to cooperate during the more lengthy procedures require careful evaluation. The overall dental health of the child, with particular reference to the caries experience, must be taken into account when making a treatment plan. The complex conservation of one tooth in the presence of a number of comparable teeth of doubtful prognosis is poor paediatric dentistry and should be avoided.

The difficulties encountered in endodontic therapy for children are related, in the main, to the morphology of the root canals and the pathological changes which may occur in primary teeth, as well as to the problems associated with incomplete root formation in the permanent dentition.

Primary dentition

Diagnosis of pulpal pathology

The reaction of pulp tissue in primary teeth to deep caries differs from that seen in the permanent dentition and is characterised by the rapid spread of inflammatory changes throughout the coronal portion of the tooth. These pathological changes become irreversible and, if left untreated, will involve the radicular tissue. There may be few, if any, clinical symptoms in the early stages to indicate the extent of tissue damage. Pain may only occur after involvement of the periradicular tissues in the spread of infection.

Children are often unable to give accurate details of their symptoms, and the responses to clinical tests may be unreliable. Difficulties are frequently experienced in ascertaining the condition of the pulp from clinical findings. Radiographs, which are essential prior to the commencement of treatment, give little information of early pathological changes.

Morphology of the primary tooth

In the primary tooth, the enamel and dentine are thinner than in the permanent tooth, and the pulp chamber, with its extended pulp horns, is relatively larger. The molars have root canals which are irregular and ribbon-like in shape, becoming narrower due to the deposition of secondary dentine. Accessory vascular canals are found in the interradicular area, where the floor of the pulp chamber is thin. Bone loss associated with an infected primary molar tooth is often interradicular rather than periapical in site, the presence of the accessory canals accounting in part for this phenomenon (fig. 1).

An additional problem is the close relationship of the roots of the primary teeth to the developing permanent successor. During exfoliation the roots of the former resorb, necessitating the use of a resorbable paste in endodontic treatment. It is also important to remember that trauma to, or infection of, a primary tooth, may result in damage to the permanent tooth. This may vary from enamel hypomineralisation and hypoplasia (fig. 2) to, more rarely, the delayed or arrested development of the tooth germ (fig. 3).

Treatment techniques

Indirect pulp capping

This technique may be of value in managing deep carious lesions, without associated radiographic changes in the periradicular tissues. The use of a local anaesthetic is essential to ensure adequate cavity preparation, in which the walls are sound and free from caries. Freedom from salivary contamination, obtained by adequate isolation of the tooth, is also important. The superficial carious dentine is carefully excavated, avoiding pulpal exposure, and the deeper layers of softened dentine are dressed with a calcium hydroxide-containing cement, eg Dycal* or Life. It is essential to seal the dressing effectively with an accelerated zinc oxide/eugenol cement, eg Kalzinol, in order to prevent contamination by saliva and bacteria. After a minimum period of 6 weeks the tooth, which should have been symptomless, is reopened, and the arrested carious lesion examined. It should appear darker and firmer. Any soft dentine remaining is carefully excav-

*Details can be found at the end of the chapter.

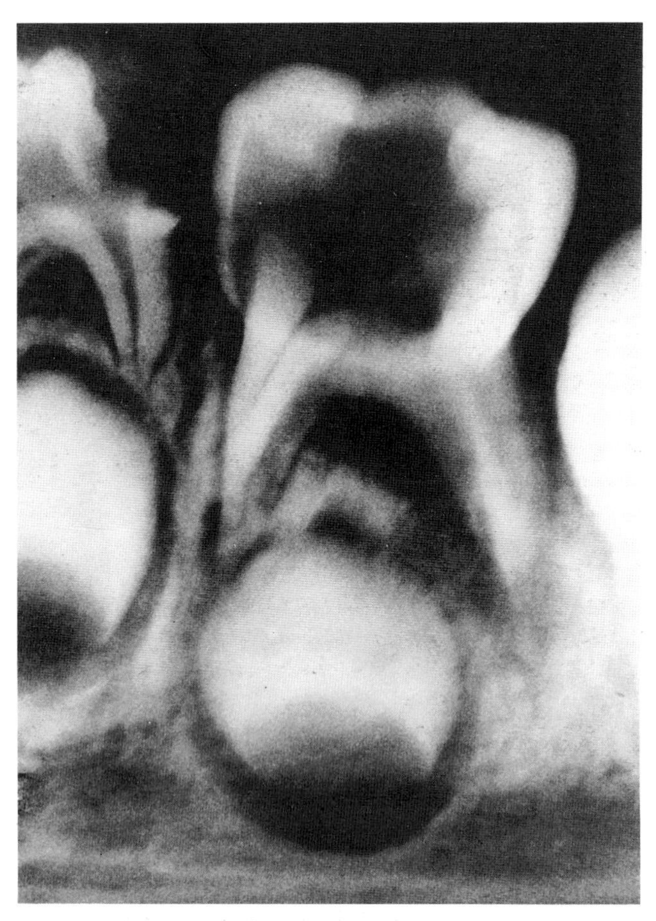

Fig. 1 Grossly carious lower second primary molar showing inter-radicular bone loss.

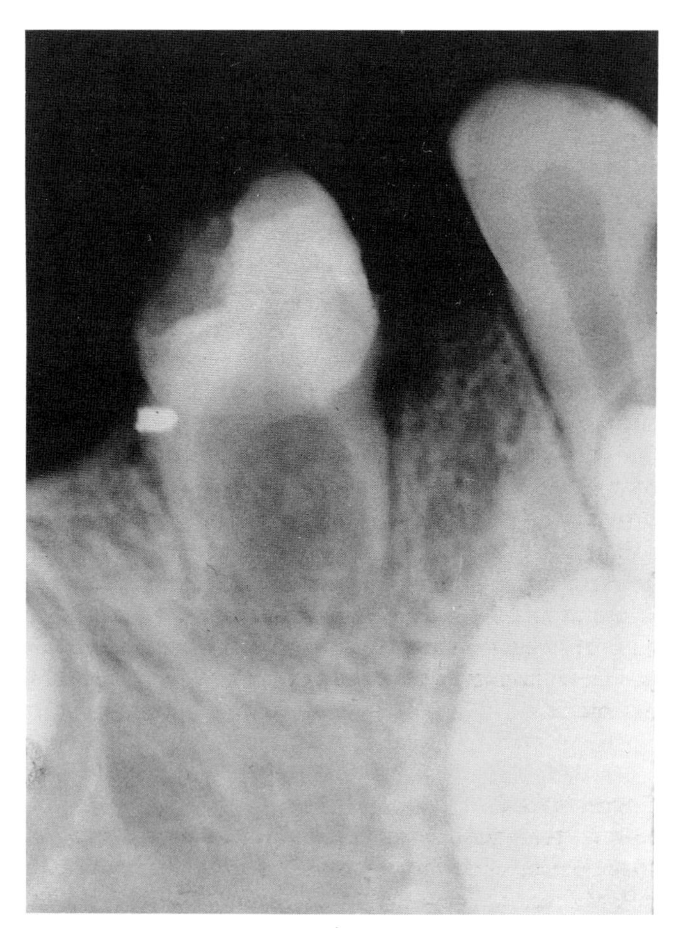

Fig. 3 Radiograph of hypoplastic premolar which has erupted prematurely, with immature root formation. The primary molar was infected.

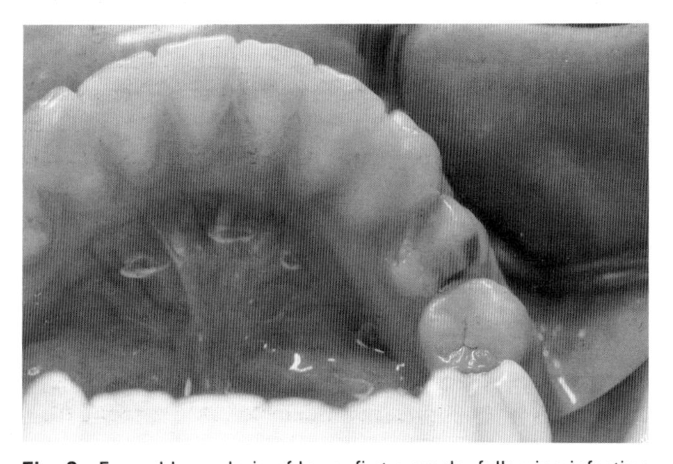

Fig. 2 Enamel hypoplasia of lower first premolar following infection of primary predecessor.

ated or removed with a large slow running bur, before a final dressing of a calcium hydroxide containing cement is placed, and an appropriate lining and final restoration inserted. For the reasons mentioned above, the success of this treatment is less predictable than in the permanent dentition. If symptoms develop after the initial stage of treatment, a devitalising pulpotomy should be employed.

Direct pulp capping

This treatment is only recommended when a small traumatic exposure occurs, during cavity preparation of a vital non-infected pulp. A calcium hydroxide dressing is placed directly over the pulp, followed by a lining and restoration, and the whole technique is carried out using local anaesthesia and with adequate isolation from salivary contamination.

Vital pulpotomy techniques

These techniques involve the removal of inflamed coronal pulp tissue and the application of a dressing to the radicular pulp in an attempt to either promote healing of, or fix, the upper portions, and to preserve the vitality of the apical tissue. Because of the difficulties involved in diagnosing the condition of the pulp tissue histologically before the commencement of treatment, careful assessment must be made at each stage of the procedure. Whenever the haemorrhage from the radicular pulp stumps is profuse and uncontrolled, the assumption is made that the inflammatory process has extended into the radicular tissue, and the therapy modified accordingly.

Calcium hydroxide

Early work[1] with calcium hydroxide used as a dressing gave disappointing results, with the majority of teeth showing areas of internal resorption which developed within a year. This resorption appears to be related to areas of inflammatory tissue which precede it. More recently, favourable results have been achieved with calcium hydroxide,[2] when it has been applied in carefully controlled circumstances. Failure of this technique is explained by the presence of an extra-pulpal clot separating the calcium hydroxide from the pulpal tissue and thus impairing healing.[3]

The treatment is carried out using local anaesthesia and adequate isolation. Following cavity preparation in the normal manner, the deep caries is removed and the coronal pulp chamber opened, such that there is no overhanging dentine inhibiting the complete removal of the pulp tissue. The coronal tissue is removed using a high-speed diamond bur cooled with sterile water or saline. Sterile cotton wool is applied to the radicular pulp tissue to achieve haemostasis and, if this is satisfactory, a dressing of calcium hydroxide, either as cement or paste, is placed over the base of the cavity. A layer of zinc oxide/eugenol cement provides a lining to the final restoration.

Both the calcium content and alkaline properties of the dressing are important to achieve healing. An initial layer of necrotic tissue develops, which becomes associated with an inflammatory reaction. Subsequently, a matrix forms and mineralises to become a hard tissue barrier of dentine-like material.

Formocresol

In this technique, Buckley's formocresol (35% tricresol, 19% formaldehyde in aqueous glycerine) is used as a pulpal dressing, in order to fix the inflamed tissue and bacteria and thus allow healing of the unaffected pulp. Small areas of internal resorption are seen in the majority of cases, but they appear to be limited in their progress.[4] However, there are problems in controlling the spread of the medicament through the radicular tissue to the periapical foramen. In addition, the toxic nature of the formaldehyde may produce adverse effects on the soft tissues through spillage or leakage. In an attempt to overcome these problems a 1:5 dilution of the solution, which is equally effective as a tissue fixative, is generally used. Recent concern led to investigations of pulpotomies employing a 2% glutaraldehyde solution as an alternative dressing. This has proved equally successful,[5] with more effective tissue fixation of the coronal portions and more vital tissue remaining in the apical portion of the canal.

The coronal pulp chamber is opened as previously described, and the pulp tissue removed with either a sharp excavator or large round bur, at slow speed, taking care that the periphery of the tooth or the floor of the chamber is not weakened or perforated. The chamber is irrigated with sterile water, or normal saline solution, to remove the debris, and gentle pressure applied with sterile cotton wool to dry the cavity and control the haemorrhage. The pulp stumps may then be assessed; if the bleeding has stopped it is assumed that normal rather than inflamed tissue is present in the canals. A cotton wool pledget moistened with formocresol solution is placed in the pulp chamber for 5 minutes, after which the pulp tissue should appear brown–black due to fixation. Zinc oxide powder, eugenol and formocresol solution is placed over the floor of the pulp chamber, followed by a further layer of an accelerated zinc oxide/eugenol cement and a final restoration (fig. 4).

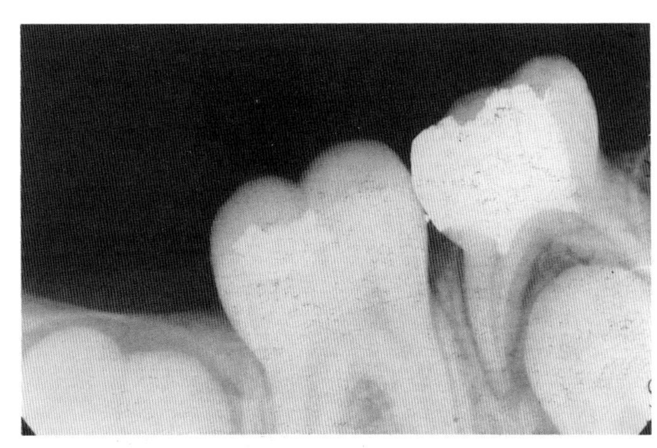

Fig. 4 Radiograph on completion of formocresol pulpotomy on lower second primary molar.

Paraformaldehyde paste

Where haemorrhage is uncontrolled before or following the application of formocresol, or endodontic procedures are restricted by inadequate anaesthesia, a 'two visit technique' may be employed. Pain prior to treatment is another possible indication for the technique. This method mummifies and fixes the coronal pulp tissue, whilst the major part of the radicular pulp remains vital, and it is therefore considered by some authorities to be more reliable.[6]

If the tooth is not anaesthetised, cavity preparation is carried out as far as possible and access is gained to the pulpal exposure. In an anaesthetised tooth, cavity preparation and removal of the coronal pulp are carried out.

A small amount of paraformaldehyde devitalising paste (paraformaldehyde 1·00 g, carbowax 1500 1·30 g, lignocaine 0·06 g, propylene glycol 0·5 g, carmine to colour) on a pledget of cotton wool is applied to the exposed pulp tissue. A soft layer of temporary dressing is then placed, without applying pressure, to seal the medicament in position; the child and parent must be warned of possible discomfort, for which analgesics are recommended. Seven to 10 days later, the tooth is checked for signs and symptoms. The devitalised coronal pulp may now be removed, without the need for local anaesthesia. A layer of zinc oxide mixed with eugenol and formocresol is then placed over the radicular stumps and the tooth restored. If some vital tissue remains in the coronal pulp chamber, a further dressing of paraformaldehyde paste is required.

Modified pulpectomy

Clinical and radiographic examination may reveal a non-vital tooth which, if not excessively mobile and without root resorption, may have a reasonable prognosis.

The cavity preparation and removal of the coronal pulp, which is necrotic, is carried out as previously described. Some workers advocate the extirpation of radicular tissue with files, following an assessment of the canal length with a pre-operative radiograph.[7] However, owing to the morphology of the root canal system and the limited access, this may not be feasible. The chamber is irrigated to remove the debris and a pledget of cotton wool moistened with Beechwood creosote, a disinfectant, is sealed in for 7–10 days. At a further visit the tooth should be symptom-free, firm, without a discharging sinus and the canals free of pus. If not, a second application of Beechwood creosote is required. A dressing of zinc oxide/eugenol, with or without the addition of formocresol, is packed into the base of the chamber and the tooth finally restored. An alternative dressing to Beechwood creosote is Kri liquid, with Kri paste used as a sub-base for the completed restoration.

Following any form of endodontic treatment, regular clinical and radiographic reviews must be made of the tooth involved and its successor. The most common change to be seen is that of internal resorption, which may occur in limited areas in formocresol pulpotomies, but may be more extensive following the use of calcium hydroxide. It may progress to cause perforation of the root.

Inflammatory follicular cysts[8] may develop, which necessitate the removal of the primary tooth and marsupialisation of the cyst to allow the permanent tooth to erupt.

Permanent dentition

First permanent molar

The first permanent molar may, soon after eruption, show extensive caries, sometimes associated with hypoplasia.

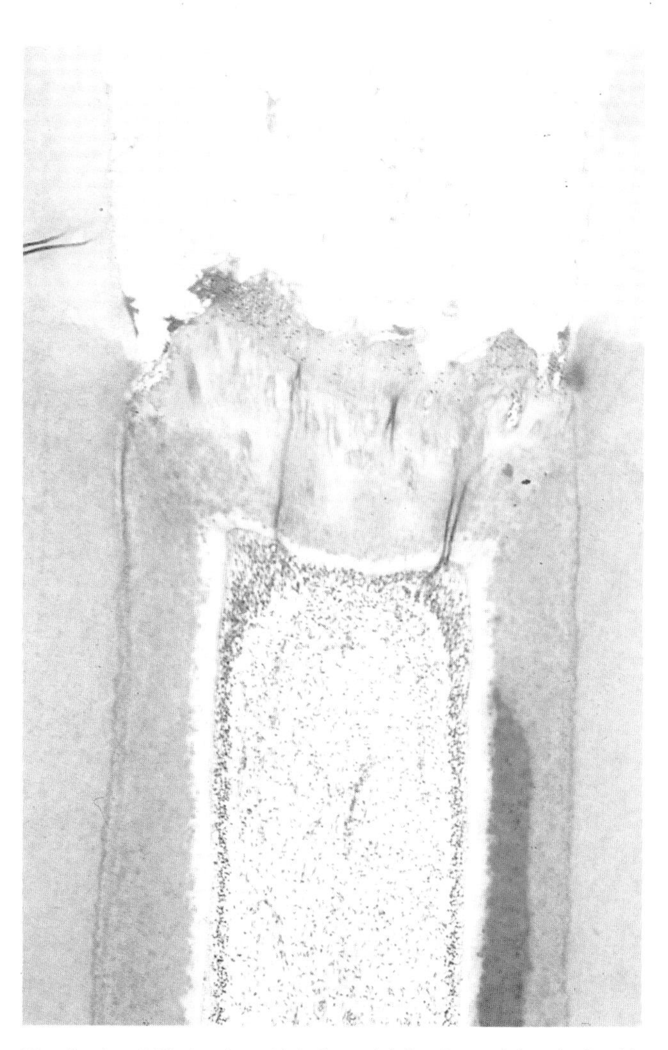

Fig. 5 A calcific barrier which formed following calcium hydroxide pulpotomy. The pulp shows no signs of inflammation and there is a well defined layer of odontoblasts.

Consideration must be given to the age of the patient and the dental development, the occlusion and possible need for orthodontic treatment, as well as the long-term restorative prognosis of the tooth and the patient's ability to tolerate involved treatment over a long period. Where necessary, planned extractions should be considered. The primary aim of conservation is to ensure that root growth continues with completion of apical formation, so that definitive endodontic treatment, if required, may be carried out at a later stage.

The vitality of the tooth must be assessed and radiographs should be available, showing the extent of carious involvement and the state of the periapical tissues.

Vital pulp

To ensure adequate preparation of the tooth, particularly in the presence of deep caries, it is essential that a local anaesthetic is administered and salivary control achieved by adequate isolation.

Indirect pulp capping

Wherever possible the technique previously described for the gradual removal of deep caries is carried out. Calcium hydroxide cement as a dressing produces more regular secondary dentine and is less irritant to the pulp tissue than zinc oxide/eugenol.[9] The tooth should be sealed effectively to prevent further bacterial contamination.

Direct pulp capping

If a small exposure of a vital tooth occurs, either accidentally during cavity preparation or because of caries, and the surrounding tissue is healthy, a direct pulp cap, with calcium hydroxide cement may be applied. A lining of reinforced zinc oxide/eugenol is then placed prior to the amalgam restoration.

Pulpotomy

If the exposure is large and the vitality of the radicular pulp is to be maintained to allow for root development, a pulpotomy is carried out. Following the opening of the coronal pulp chamber and the removal of the pulp tissue, the area is irrigated and dried. Haemostasis of the radicular pulp should be observed prior to the application of calcium hydroxide cement or paste, and the provision of a permanent restoration. A calcific barrier should develop adjacent to the dressing (fig. 5), and root development continue in the presence of healthy pulp tissue.

Non-vital pulp

Endodontic treatment of a non-vital, young, permanent tooth should be undertaken only after careful assessment of the developing occlusion, the condition of the comparable teeth, the patient's ability to cooperate and the long-term prognosis of the tooth.

If pulpal necrosis occurs prior to the complete development of the apex, the objective of treatment is to encourage further deposition of calcified tissue in the apical region. Thorough preparation of the root canals is carried out, avoiding damage to the apical tissues and cells of Hertwig's root sheath. Calcium hydroxide is then applied, as previously described in the section on calcium hydroxide. Definitive endodontic treatment is carried out when an apical stop has formed and the tooth is then restored.

In fully developed, young, permanent teeth root-filling with gutta-percha and sealer is indicated.

References

1 Magnusson B O. Attempts to predict prognosis of pulpotomy in primary molars. *Scand J Dent Res* 1970; **78:** 232–240.
2 Schroder U. Two year follow up of primary molars pulpotomised with gentle technique and capped with calcium hydroxide. *Scand J Dent Res* 1978; **86:** 273–278.
3 Schroder U, Granath L. On internal dentine resorption in deciduous molars treated by pulpotomy and capped with calcium hydroxide. *Odont Revy* 1971; **22:** 179–188.
4 Magnusson B O. Therapeutic pulpotomies in primary molars with the formocresol technique. *Acta Odont Scand* 1978; **36:** 157–165.
5 Garcia-Godoy F. Clinical evaluation of glutaraldehyde pulpotomies in primary teeth. *Acta Odontol Pediatr* 1983; **4:** 41–44.
6 Hobson P. Pulp treatment of deciduous teeth. *Br Dent J* 1970 **128:** 275–282.
7 Rifkin A. A simple effective safe technique for the root canal treatment of abscessed primary teeth. *J Dent Child* 1980; **47:** 435–441.
8 Shaw W, Smith D M H, Hill F J. Inflammatory follicular cysts. *J Dent Child* 1980; **47:** 97–101.
9 Glass R L, Zander H A. Pulp healing. *J Dent Res* 1949; **28:** 97–107.

Dycal: L. D. Caulk Co., Delaware, USA.
Life: Sybron/Kerr, Michigan, USA.
Kalzinol: De Trey, Weybridge, England.
Kri: Pharmachemie, Zurich, Switzerland.
Formocresol and paraformaldehyde devitalising pastes: available as a *SPECIAL*, from Macarthys Laboratories Ltd., Chesham Close, Romford, RM1 4JX.
Beechwood creosote is also available from Macarthys Laboratories Ltd.

10

Surgical Endodontics

Conventional orthograde root canal therapy is the preferred method of treating the irreversibly damaged pulp. However, there are occasions when a surgical approach may be necessary, because access to the root canal system and the affected area is not possible using non-surgical techniques. Surgery may be necessary to establish drainage, to seal the contents of the canal system from the surrounding tissues, or to repair any defects in the tooth root. Surgical resection of a multi-rooted tooth may be required where, for technical reasons, one of the roots cannot be successfully treated. The general indications for endodontic surgery have been well documented.[1–7] The following procedures will be discussed: incision to establish drainage, periapical curettage, apicectomy, surgical repair of roots, root amputation, and hemisection.

Medical and dental considerations

A well-documented medical history is essential (*see* Chapter 2). In general, heart disease, diabetes, blood dyscrasias, debilitating illnesses and steroid therapy may contra-indicate surgery and special measures are necessary if surgery is contemplated. Consideration must also be given to psychological factors. As a rule, local analgesia is preferable, but patients who are particularly apprehensive may wish to have any surgery carried out under sedation or general anaesthesia. The choice of anaesthetic will also be governed by the nature of the operation, the site of the tooth and ease of access. A history of rheumatic fever is not a contra-indication for endodontic surgery, provided antibiotic cover is given. If there is any doubt about a patient's fitness to undergo any surgical endodontic procedure, then the patient's physician should always be consulted.

The first considerations are whether the tooth is worth saving and how important it is in the overall treatment plan. An assessment must also be made of the effects of any proposed surgery on the periodontal condition. The presence of any detectable dehiscence or bony fenestration will influence the design and extent of the flap.

Good visual access is extremely important, and the anatomy of the area must be thoroughly understood. The position of any major structures such as neurovascular bundles and the maxillary sinus must be noted. Assessment should also be made of the root shape, taking into account any unusual curvature and the number of foramina that may be exposed at the apex as a consequence of the operation.[5] A buccal or labial approach is always preferred, as a palatal approach is difficult and should only be undertaken in exceptional circumstances by experienced practitioners.

Incision to establish drainage

Incision to establish drainage is the only surgical endodontic procedure which may be undertaken when acute inflammation is present. The principal indication is the presence of a collection of pus which points from a fluctuant abscess in the soft tissues. Establishing drainage to help bring the infection under control is essential, and should be obtained through the root canal and soft tissues in preference to administering antibiotics alone.[8] Examine the soft tissue swelling to see if it is fluctuant. Where the swelling is pointing intra-orally, apply copious amounts of surface analgesia, for example ethyl chloride or topical lignocaine ointment. Sometimes, a few drops of local analgesic solution can be infiltrated submucosally, but there is a risk of spreading the infection further into the tissues and this has to be considered. Incise the swelling with a Bard-Parker No. 11 or 15 scalpel blade, or aspirate, using a wide bore needle and disposable syringe.[9] It may be possible to aspirate the abscess via the root canal as well. The advantage of this technique is that the sample can be sent for bacteriological examination if required. It is not usually necessary to insert a drain, but if it is thought necessary then a piece of quarter-inch or half-inch selvedge gauze may be used.[10] The same criteria apply when extra-oral drainage is indicated, and it may be possible to use the same technique of aspiration with a wide bore needle and disposable syringe.[9] If an extra-oral incision is considered necessary, it is wise to refer the patient to an oral surgeon for this particular procedure.

Patients may present with an acute apical abscess that is still confined to cancellous bone. The tooth is exquisitely tender and it may not be possible to establish drainage through the root canal, because the tooth has a post crown, for example. Under these circumstances, it may be tempting to try 'trephination' (surgical fistulation). This is a difficult procedure and the technique cannot be recommended. The apex is not easy to locate under such circumstances and the root itself may be easily damaged (figs 1–5) and will require a more complex operation to repair the damage.

Periapical curettage

The object of this procedure is to remove any soft tissue lesion from around the root apex and to remove any necrotic cementum. This used to be a routine operation carried out by many practitioners after completion of a root canal filling. The rationale for this is no longer accepted, because if the root filling has been carried out successfully and the canal system has been sealed, then

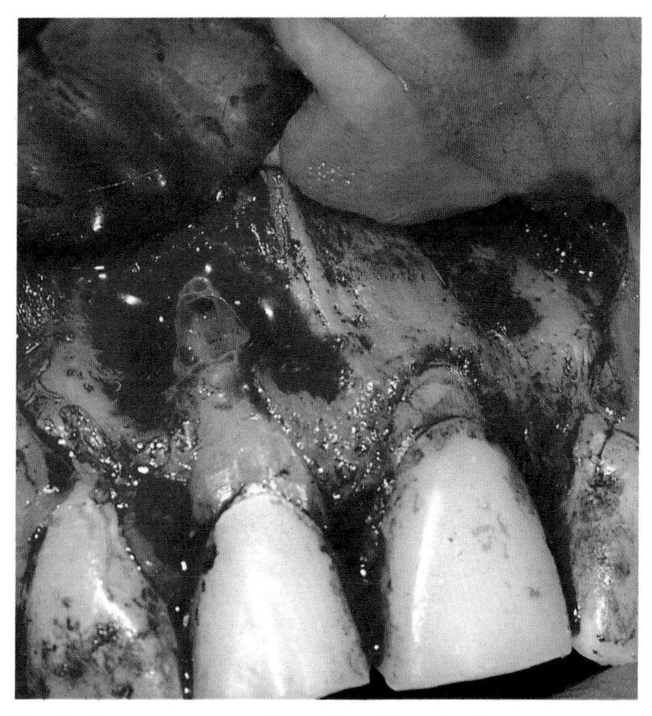

Fig. 1 An attempt was made to relieve this patient's acute symptoms using 'trephination', with an unfortunate result. The apex has been missed and the root severely damaged by the procedure.

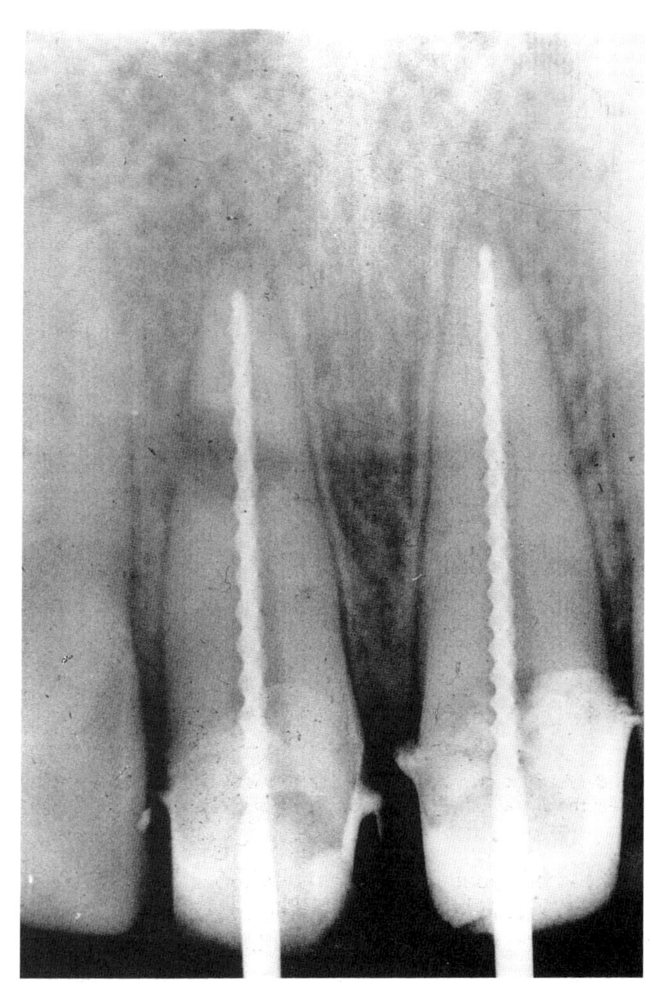

Fig. 2 Both teeth were prepared using conventional orthograde techniques. The root canal in the 1|(11) was filled with calcium hydroxide to dry out the canal and to allow time for the tissues to recover.

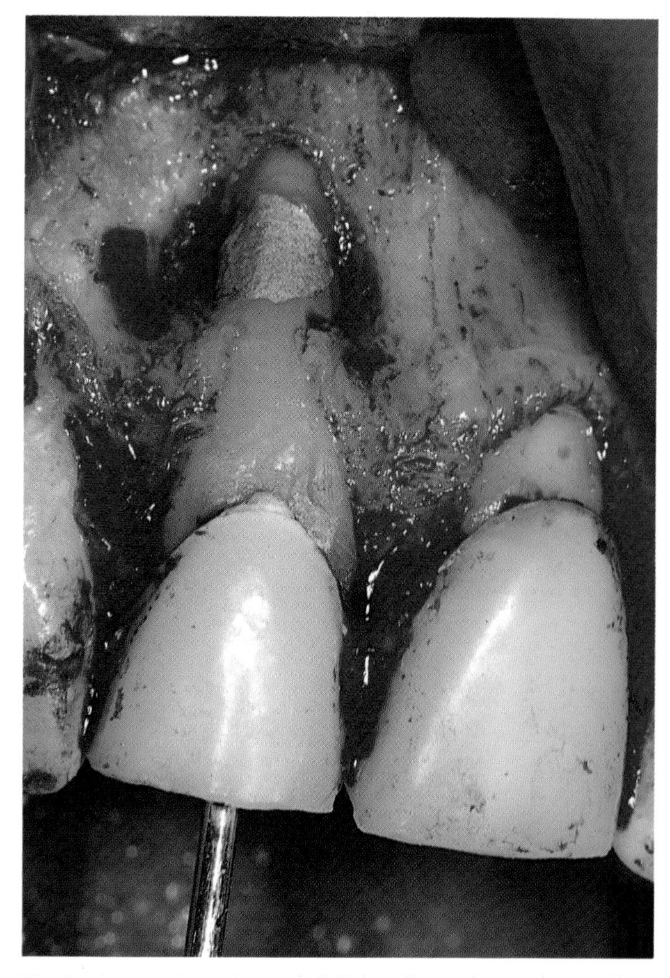

Fig. 3 At operation, a large, well-fitting silver point was inserted into the canal and the root defect repaired with amalgam. The silver point was gently rotated to ensure that it could be removed from the canal once the amalgam had set.

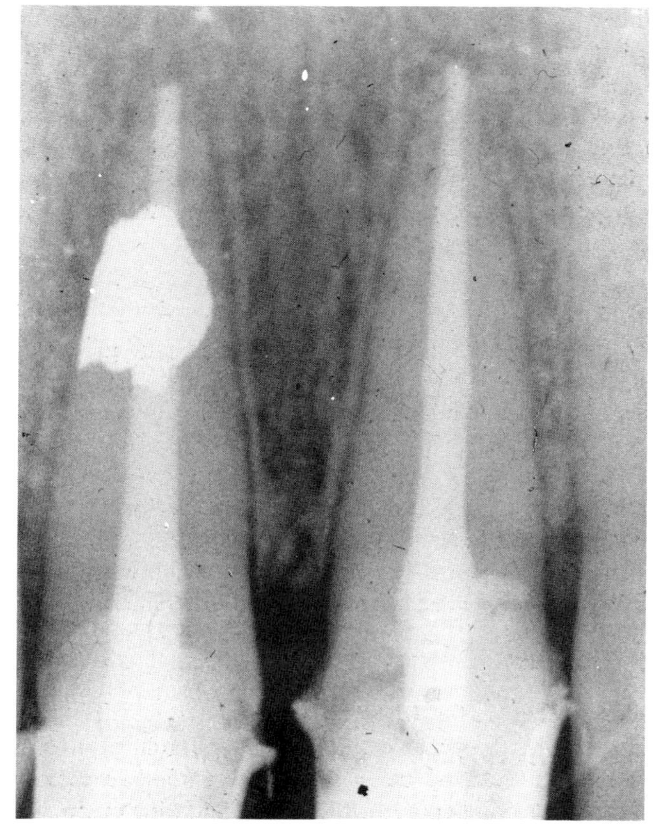

Fig. 4 The completed root canal fillings in 1|1 (11, 21).

healing of the lesion will take place without surgical intervention. Periapical surgery is only necessary if the periapical lesion fails to respond to conventional orthograde root canal treatment. When undertaking periapical surgery, as much as possible of the periapical lesion should be removed, although it is not thought necessary to remove it all. The soft tissues in a periapical lesion are essentially healing and defensive in nature and some practitioners do not consider it desirable to remove all of the soft tissue from the area.[14] This is fortunate as, technically, it is difficult to remove every trace of the lesion, especially if it is firmly attached to the wall of the bone cavity.

At this stage, it would seem reasonable to proceed with an apicectomy and retrograde root filling to ensure that the exiting canal at the root apex is sealed. However, if there is a well condensed root filling and the apical seal cannot be improved upon, then this could be considered a justification for periapical curettage; the procedure is otherwise a preliminary step for carrying out an apicectomy.

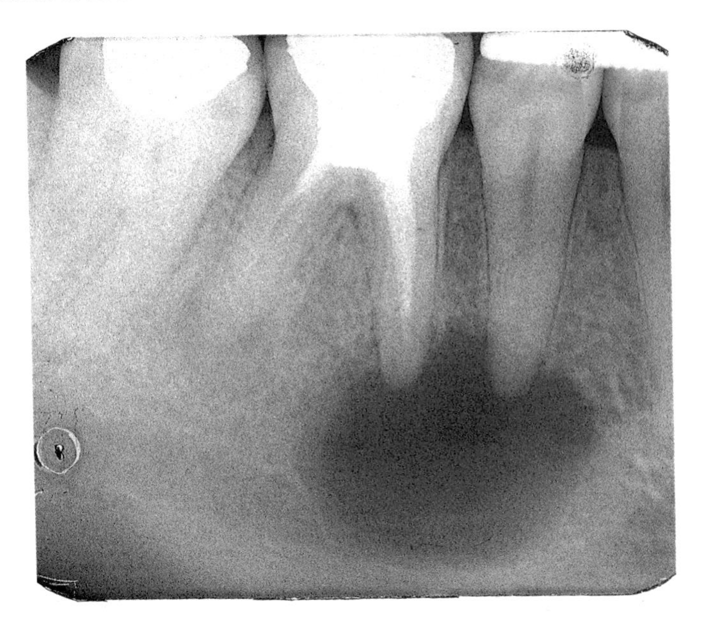

Fig. 5 Biopsy a periapical lesion when, as in this case, there is uncertainty about its nature.

Biopsy of a periapical lesion

The one specific indication for endodontic surgery is uncertainty about the nature of the apical lesion. The lesion should be excised in its entirety and sent for evaluation (fig. 5).[1–3]

Apicectomy

The term apicectomy refers to a stage of the operation only. The principal objective is to seal the canal system at the apical foramen from the periradicular tissues.[5] To do this it is necessary to resect the apical part of the root to gain access to the root canal, hence the term. An apicectomy is an adjunct measure to orthograde root treatment for two reasons. First, there is very little chance of being able to seal all the lateral communications between the canal and the periodontal ligament with a retrograde root filling technique (fig. 6). Secondly, the area of root-filling material exposed will be greater and the long-term success affected, because all root-filling materials are, to some extent, irritant to the tissues.

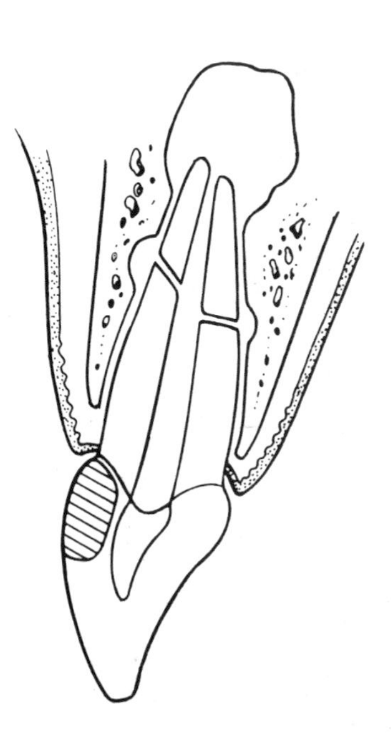

Fig. 6 An apicectomy procedure has very little chance of being able to seal the lateral canals, because of their inaccessibility.

Indications for an apicectomy

Improvements in root canal treatment techniques have lessened the need for apical surgery. Cases which at first seem obvious candidates for endodontic surgery may respond to conventional treatment provided careful thought is given to the aetiology (figs 7 and 8). Once the decision has been made to carry out an apicectomy, consideration must be given to the chances of success. Access and control of the operating environment are essential, otherwise the end result will be counterproductive (figs 9 and 10).

Retreatment of a failed root filling

An apicectomy may be considered if a root filling fails and retreatment cannot be effected by orthograde means. There are a number of reasons why a root filling might fail, but generally this is due to inadequate cleansing and filling of the root canal.[11] Some root fillings can prove very difficult to remove, for example apical silver points. Other root-filling materials are of such a hard consistency that removal of them is impractical. Occasionally there may be an anatomical reason for the failure, such as an unfilled apical delta. Attempts at retreatment by a conventional orthograde route may be unsuccessful because the original

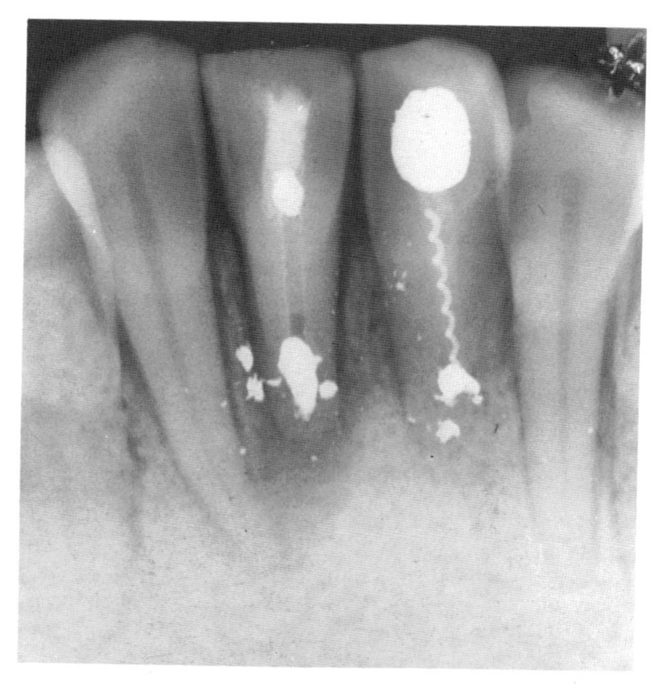

Fig. 7 At first sight there would seem to be little hope for the mandibular central incisors. A second attempt at surgery will fail because of the poor crown–root ratio. The chances of being able to seal the canals apically using a surgical approach are limited.

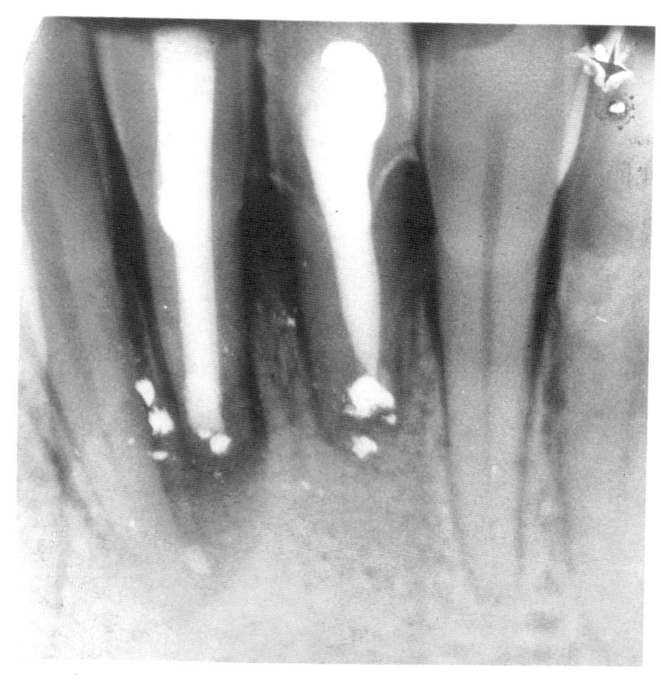

Fig. 8 The canals were meticulously cleaned, prepared and filled using the conventional orthograde route.

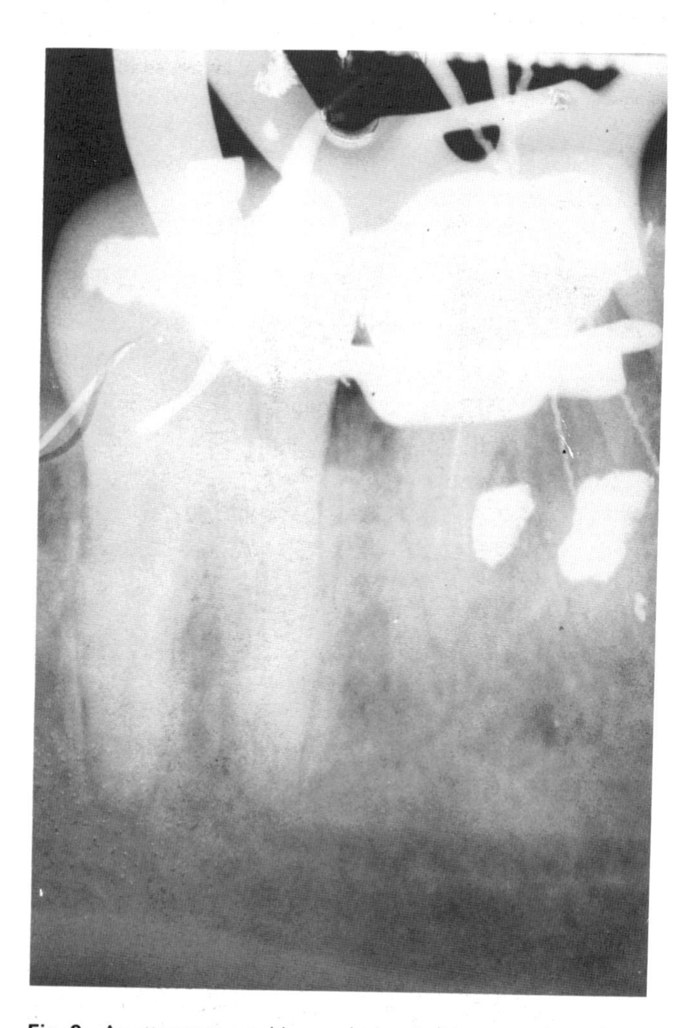

Fig. 9 An attempt to provide an apical seal with retrograde amalgams has failed. The amalgam seals are embedded in the surrounding bone and nowhere near the apex. Good visual access and control of instrumentation is essential. The mandibular molar region is a difficult area to carry out an apicectomy and should be avoided.

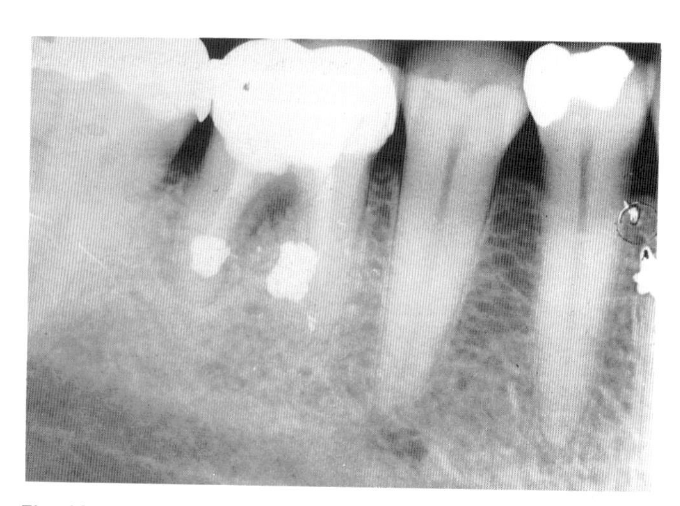

Fig. 10 The mandibular first molar was treated successfully by the orthograde route.

canal cannot be negotiated (figs 11–13). An apicectomy is therefore justified to ensure that the apical foramen has been sealed.

Root-filling material which has been extruded through the apical foramen may be a contributory cause of failure, since it could be an indication that the apical seal is deficient; necrotic material may be present at the apex and between the interface of the root filling and the canal wall; the root-filling material itself may be highly irritant (figs 14 and 15).

Limited time for treatment

Treatment may have to be carried out at one visit as the patient cannot spare the time for a series of appointments. If there is a periapical lesion present, or some of the problems already mentioned are likely to arise, these could be considered valid indications for endodontic surgery.

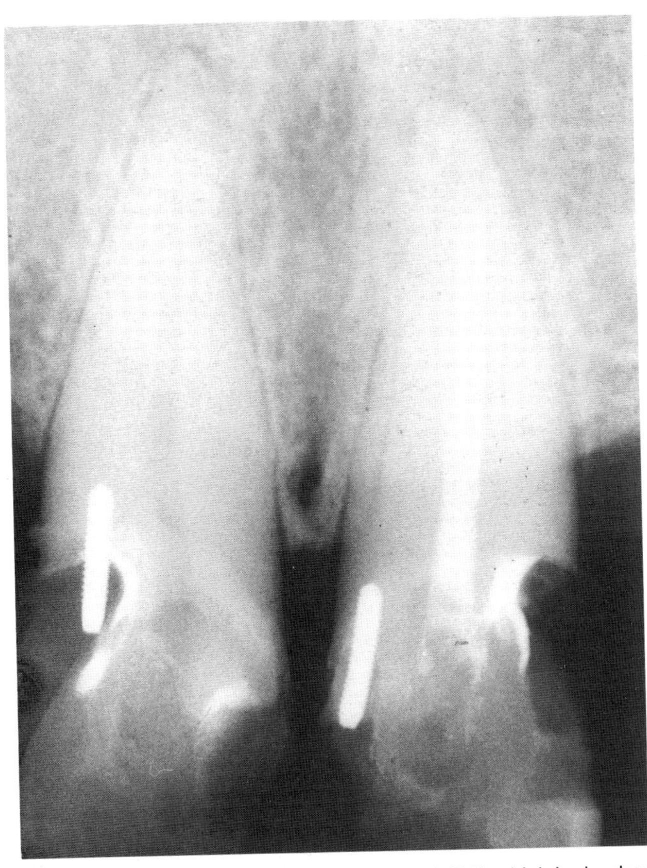

Fig. 11 The treatment was required for the |1 (21) which had a short root filling.

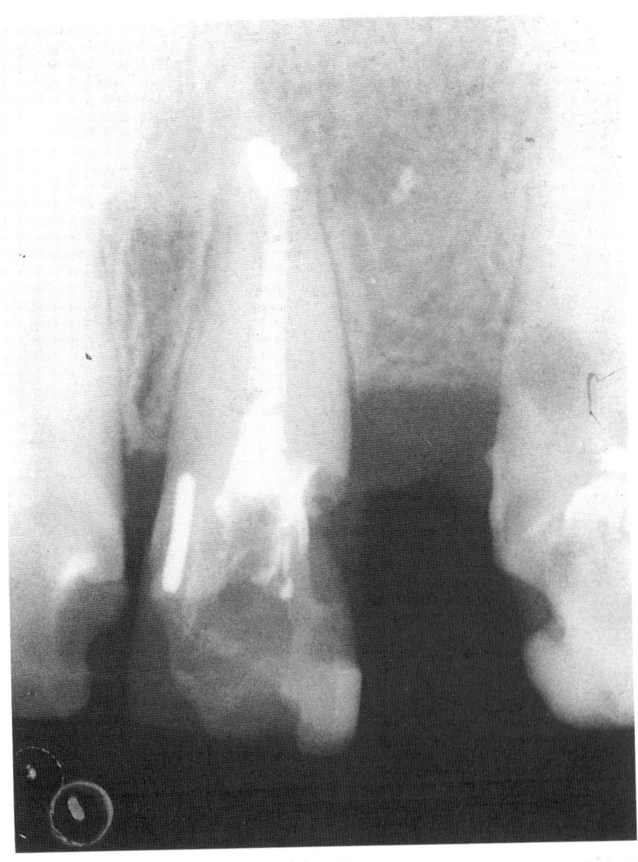

Fig. 13 The post-operative radiograph of the completed root canal filling and retrograde amalgam.

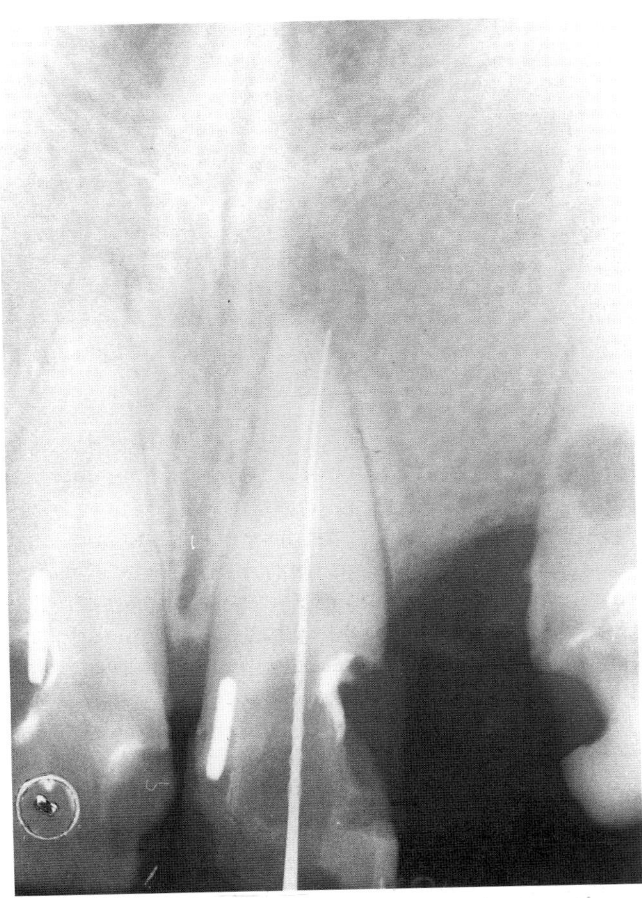

Fig. 12 It proved impossible to negotiate the original canal, so an apicectomy was carried out to seal the original canal and the small perforation beside it.

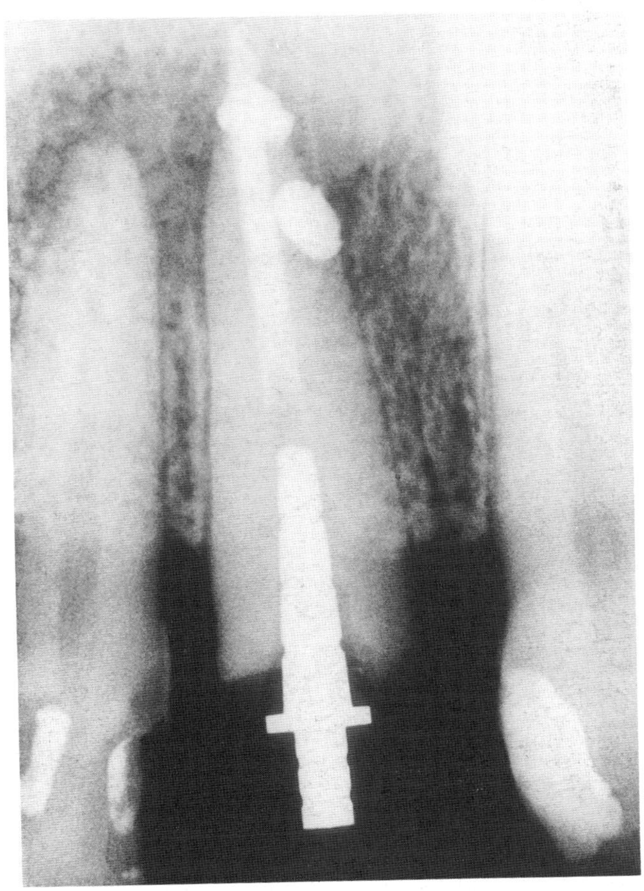

Fig. 14 Excess material through the apex was causing symptoms. An apicectomy is necessary to remove the extruded material and provide an apical seal.

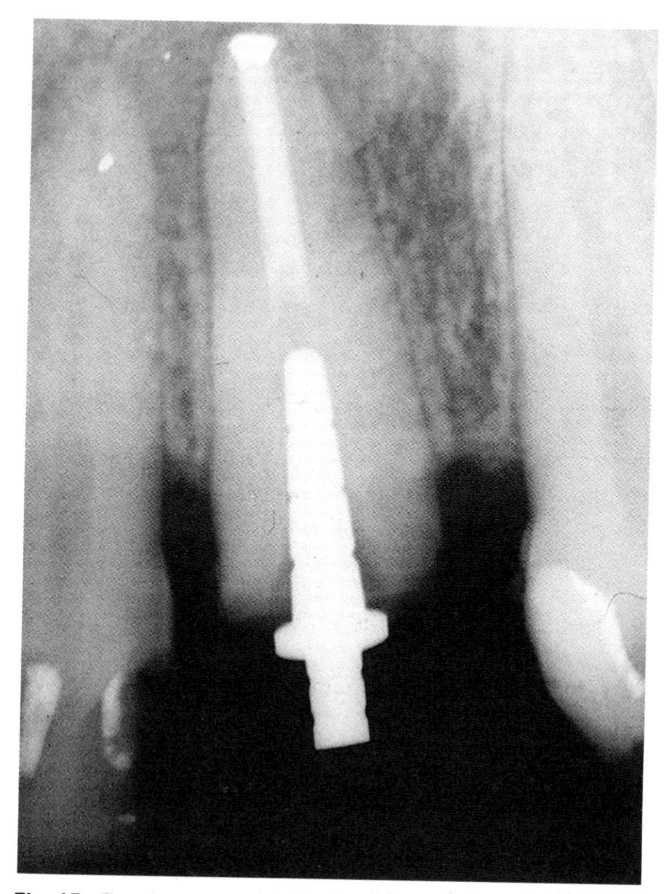

Fig. 15 The post-operative radiograph showing excess material and debris removed and a retrograde amalgam neatly in place. (Surgery by Miss E. Watts, Eastman Dental Hospital.)

Procedural difficulties
During conventional orthograde root canal treatment, problems may arise as a result of one of the following:

(1) Unusual root canal configurations.
(2) Extensive secondary dentine deposition.
(3) Fractured instruments within the root canal.
(4) Open apex.
(5) Existing post in the root canal.

Unusual root canal configurations
Instrumentation of canals in roots which exhibit severe dilaceration may prove impossible. Similarly, where there is an apical delta, thorough cleansing, shaping and obturation of the canal may be impractical and surgery will be required to complement the orthograde approach.

Extensive secondary dentine deposition
The ageing process results in the deposition of secondary dentine, with a consequent reduction in size of the pulp chamber and the root canal. The canals are sometimes completely obliterated and it may prove impossible to negotiate the canals with even the smallest size instruments. Under these circumstances, apical surgery is the only alternative to extraction.

Fractured instruments
Instruments which have fractured in a root canal do not necessarily result in failure of the root treatment. They should be removed if possible, but if this is impossible, then an attempt should be made to seal the rest of the canal with the instrument in place. Surgery is only necessary if the tooth develops symptoms.

Open apex
Teeth that have been injured before root development is complete should be treated conventionally in the first instance. If the pulp is vital, then the coronal pulp is removed and the remaining vital radicular pulp is covered with calcium hydroxide to allow continued root development; this is termed 'apexogenesis'. If the pulp is irreversibly inflamed, then the radicular pulp is removed and the canal filled with calcium hydroxide to encourage root formation and closure of the apex; this is termed 'apexification'.[12] However, should the treatment fail, then apical surgery is indicated, to provide an apical seal after completion of the orthograde root filling. Care must be exercised when carrying out this operation as the root structure is often very delicate.

Existing post in the root canal
An apicectomy may be indicated for teeth with symptomatic periapical lesions which have satisfactory post crowns in place, provided the root filling in the main body of the canal is satisfactory. However, it has to be remembered that success depends on the canal system being completely sealed and if there is any doubt about this, it is better to remove the crown and post and carry out orthograde root treatment, avoid surgery, and thus provide a sound foundation for any subsequent restoration.

Surgical repair of roots
Surgery may be necessary to repair defects in a root surface due to either iatrogenic or pathological causes. The two main indications are as follows.

Perforations
An orthograde approach should first be used to seal the perforation. The canal must be thoroughly cleaned and filled with calcium hydroxide paste (calcium hydroxide powder BP and local anaesthetic solution) to dry it out and to allow the tissues time to heal.[13]

The prepared canal space is then obturated using conventional root canal filling techniques. Perforations caused by engine reamers can usually be treated by an orthograde approach as the access is generally good. However, if clinical symptoms persist and there is bone resorption, surgery will permit placement of a better seal.[8]

Internal and external root resorption
Internal resorption should always be treated by an orthograde route first. If the resorptive process has perforated through to the periodontal ligament, then surgery may be necessary to repair the root and provide an effective seal. Certain types of external root resorption in the early stages can be dealt with by surgery, provided access can be gained to the area (*see* Root resorption).

Root amputation and hemisection
An apicectomy on a posterior tooth is a more difficult

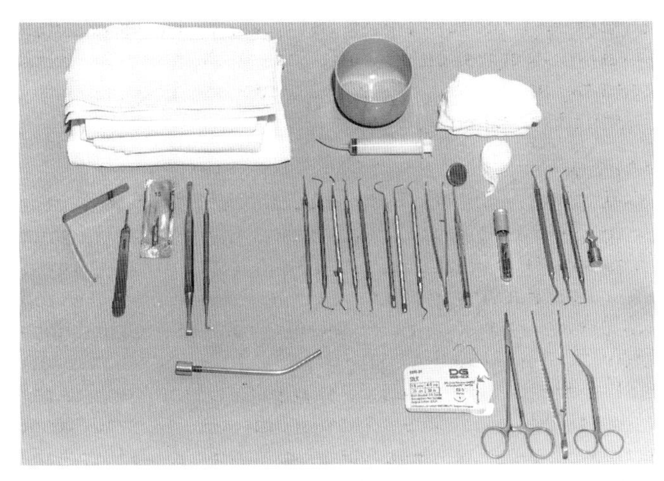

Fig. 16 (*Top row*) Sterile towels; plastic bowl for saline; Monoject 20 ml syringe; gauze swabs; gauze ribbon for packing bone cavity. (*Middle row*) Austin retractor; Scalpel handle; No. 15 Bard-Parker scalpel blades; periosteal elevator; Osteo-Mitchell trimmer; excavators; 121/213, 125/126, 206/207, G3, G5.
Probes: sickle no. 54; right-angled no. 6; Briault no. 11.
College tweezers: mirror no. 4.
Burs: Latch grip round: R 1/4, R 1/2, R 1, R5. Latch grip tapered fissure: 701, 702.
Plastic instruments: carver 156; amalgam plugger 15; burnisher 49.
Hill amalgam carrier.
(*Bottom row*) Surgical suction tip; siliconised 4·0 black silk suture; 16 mm needle; needle holders; rat-toothed tissue forceps; suture scissors.
Also required but not shown: physiological saline; self-aspirating syringe; local anaesthetic solution containing 1:80 000 adrenaline; needles 27G and 30G for anaesthetic syringe; right-angled handpieces: miniature head, standard head; straight handpiece.

procedure to carry out than on an anterior tooth. For this reason, the relatively simpler techniques of root amputation or hemisection may be considered. The changes in endodontic and periodontal treatment techniques in recent years have greatly improved the prognosis for this form of treatment. The principal indications are endodontic, restorative, or periodontal. Root amputation is an operation where the entire root of a multi-rooted tooth is removed, leaving the crown intact. Hemisection is the division of a tooth, usually in a buccolingual plane. Normally, one half of the tooth is removed, but both sections may be retained if there is disease in the furcation area only. However, the restorative problems this type of treatment poses are considerable and for this reason the prognosis is generally poor. Pre-operative assessment of both the periodontal and restorative aspects is crucial if these methods of treatment are contemplated.

Armamentarium

For all surgical procedures, instruments should be set out, preferably in the order in which they will be used. A typical layout, as used in the Eastman Dental Hospital, is shown in figure 16.

Access and flap design

The design of a flap should permit an unobstructed view of the operating area and permit easy access for instrumentation. The following points need to be considered:

(1) The blood supply to the flap and adjacent tissues must be sufficient to prevent tissue necrosis when it is repositioned.
(2) The edges of the flap should lie over sound bone and not cross any void; otherwise breakdown may occur and defective healing will result.
(3) Incisions should not cross any bony eminence, for example the canine eminence, as healing will be poor, particularly if there is a dehiscence or fenestration present.
(4) The incision must be clean, so that the flap can be reflected without any tearing of the margins.
(5) The periodontal tissues should be healthy, as healing will be affected by any overt disease.

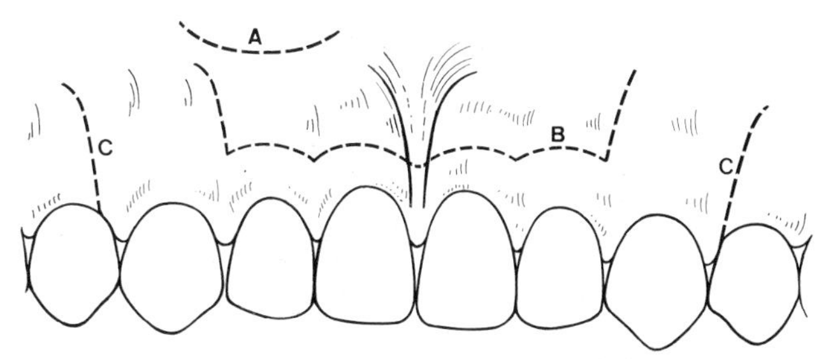

Fig. 17 (A) semilunar flap. (B) Leubke-Ochsenbein flap. (C) Full mucoperiosteal flap.

There are several designs of flaps; the three most common are the semilunar flap, the Leubke-Ochsenbein flap, and the full mucoperiosteal flap.

Semilunar flap (fig. 17a)

The incision is drawn in a semicircle from near the apex of the adjacent tooth towards the gingival margin around the area to be operated on, and finishes near the apex of the tooth on the other side. The margin of the flap should extend into the attached gingivae. Its advantage is that it is simple to replace and easy to suture. A disadvantage is the scarring which invariably accompanies this design. The principal difficulty occurs if the margins of the bony cavity extend across the incision line because the lesion is much larger than was originally apparent.

Leubke-Ochsenbein flap (fig. 17b)

The Leubke-Ochsenbein flap has been designed to overcome some of the disadvantages of the semilunar flap. A vertical incision is made down the distal aspect of the adjacent tooth to a point about 4·0 mm short of the gingival margin. The horizontal incision is scalloped following the contour of the gingival margin through the attached gingivae to the distal aspect of the tooth on the other side. The incision must always be extended to the other side of the fraenum and the distal aspect of the adjacent maxillary central or lateral incisor to avoid a vertical incision next to the fraenum.

The flap affords an excellent view of the operating area. However, it still has the disadvantage that the margins of the bony cavity might extend across the incision line, as can happen with the semilunar flap. It is essential to check if there is any pocketing, as breakdown will be inevitable. The aim of this flap design is to preserve the integrity of the gingival margins if there are crowns on the teeth. Scarring is still a problem with this type of flap.

Full mucoperiosteal flap (fig. 17c)
This design of flap offers all the advantages of the mid-level flaps and relatively few disadvantages. Incision requires more accuracy and it takes more time to reflect the flap. Similarly, repositioning the flap is time-consuming and suturing has to be done carefully. The advantage is that it provides excellent access to the operating area, with none of the risks associated with other designs. Scarring is rarely a problem. There are several variations: the trapezoidal flap, the triangular flap, and the gingival flap with no relieving incisions. It is recommended that this design of flap is used whenever possible and is the one described in the following section.

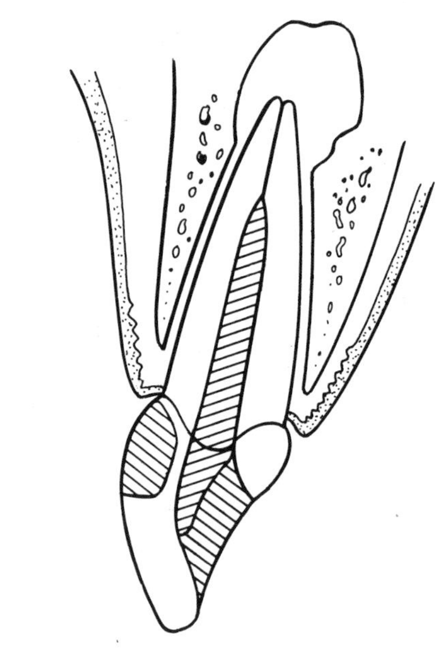
Fig. 18 Pre-operative state.

Apicectomy technique
The steps for carrying out an apicectomy are:

(1) Analgesia.
(2) Reflection of flap.
(3) Location of apex.
(4) Curettage of area.
(5) Resection of root.
(6) Retrograde cavity preparation.
(7) Retrograde filling.
(8) Replacement of flap and suturing.

Analgesia
Local analgesic solutions containing 1:80 000 adrenaline should be used to give adequate haemostasis. In the mandible, block injections should be given, in addition to infiltration of the tissues in the operating area. In the maxilla, the palate must be well infiltrated to anaesthetise the greater palatine nerve. The incisive papilla and canal must also receive sufficient anaesthetic solution to block the long sphenopalatine nerve. The best method is by gradual infiltration from the labial aspect; otherwise it will be very painful. Local anaesthetic solutions containing Octapressin do not give adequate haemostasis and should be avoided if possible.

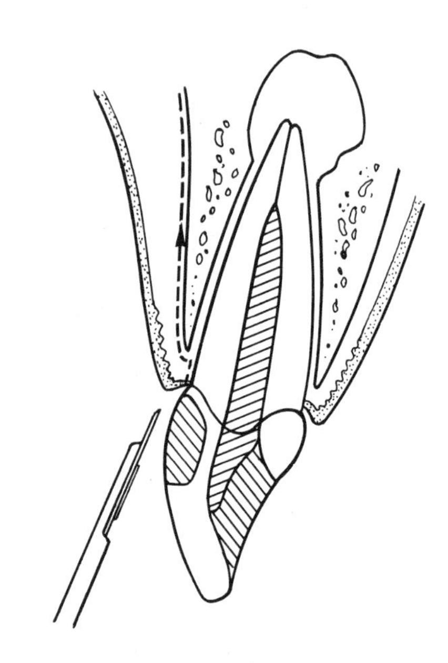
Fig. 19 Incision directed into gingival crevice.

Reflection of flap
Vertical relieving incisions are made firmly down the line angle of the teeth on either side of the operating area, into the gingival crevice, taking in the papilla. The horizontal incision is made along the gingival crevice to join vertical relieving incisions. The flap is then reflected with a periosteal elevator lifting the periosteum with it from the bone. Once freed, the flap is held in position with an Austin retractor (figs 18–20).

Location of the apex
If the lesion has perforated the cortical plate, then location is a fairly simple matter. However, if this is not the case, then measurement of the tooth from the radiograph taken with a long cone paralleling technique must be made. Initially, a size R1 latch grip round bur may be used to provide small shallow exploratory holes to locate the site of the apex and the lesion. This must be done very carefully, to avoid damaging the root surfaces of the teeth in the immediate area (figs 21 and 22). Alternatively, a size R5 latch grip round bur may be used to locate the apex by

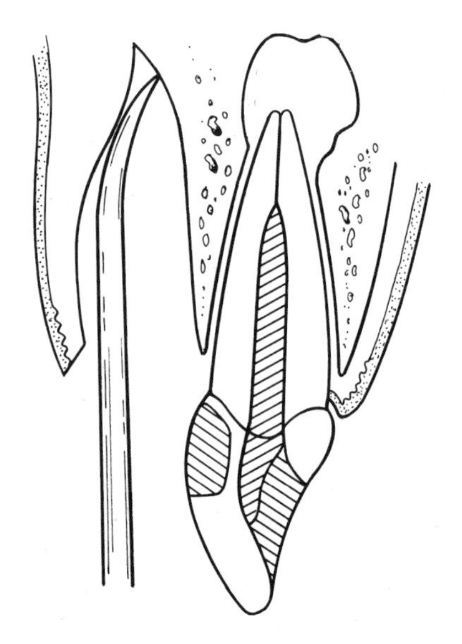
Fig. 20 Flap raised using a periosteal elevator.

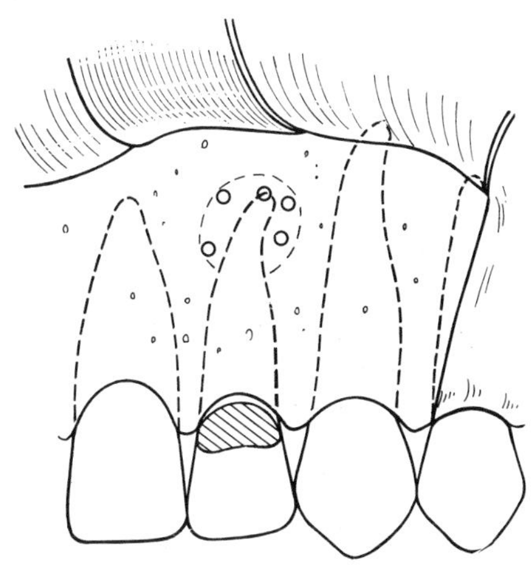

Fig. 21 Size R1 round bur is carefully used to locate apex and site of lesion.

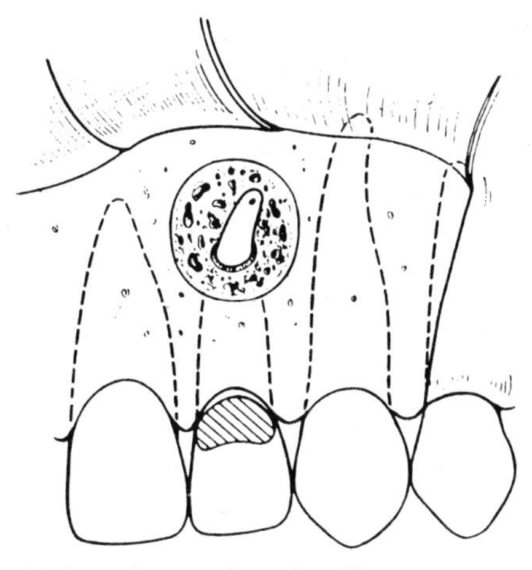

Fig. 23 Labial view of apex prior to bevelling.

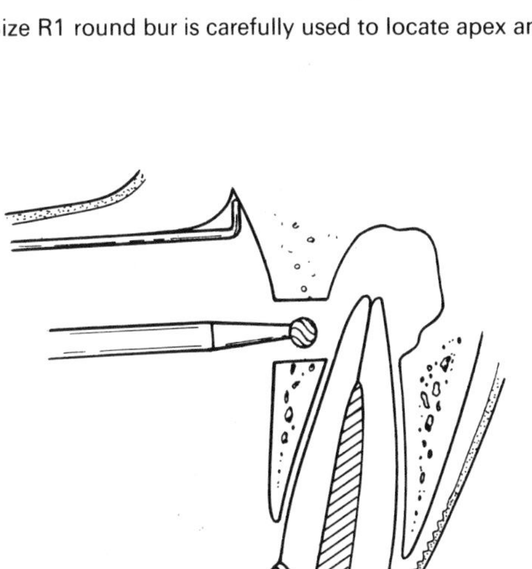

Fig. 22 Size R5 round bur is used to pare away the bone until good access to the apex is obtained.

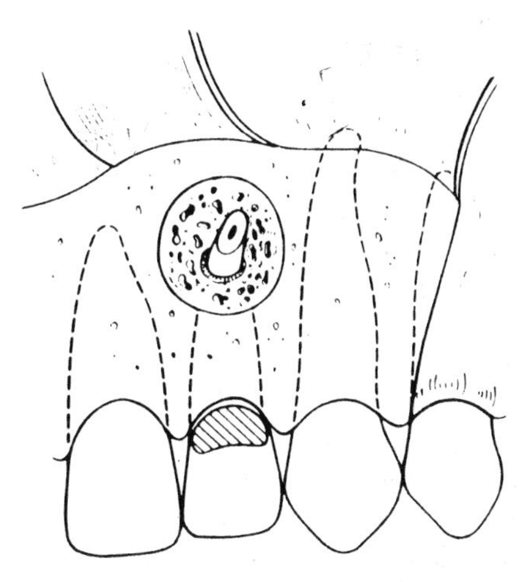

Fig. 24 Apex bevelled and minimal amount of tooth root removed.

paring away the cortical bone over the apex. More bone is shaved away using the R5 latch grip bur, until good visual access is obtained.

Curettage

Any soft tissue lesion must be carefully curetted from around the apex. It is useful to infiltrate some local anaesthetic into the lesion at this stage, as the exposed area is occasionally sensitive. Additional haemostasis is also obtained, which makes the operation easier to perform.[3] As much of the lesion as possible should be shelled out in one piece. If the tissue is firmly attached, it can be left in place and removed after completion of the retrograde filling, otherwise persistent haemorrhage may result. Some practitioners leave these tissue tags in place.[3,4]

Resection of the root

The aim of resection is to present the surface of the root so that the apical limit of the canal can be visually examined and to provide access for retrograde cavity preparation. It is not necessary to resect the apex to the base of the bony cavity. If too much root is removed, then a greater cross-section of the canal will be revealed, exposing a larger area of filling material to the tissues, and thus reducing the chances of successful healing. The amount of available root length has to be considered for any future post crown construction. There is also an inherent disadvantage as the crown–root ratio is reduced, which may affect the adaptive response of the periodontal ligament to excessive occlusal forces (figs 23 and 24).

A latch grip tapered fissure bur (701 to 702) is used to bevel the root surface at approximately 45° to the long axis of the tooth, with the bevelled surface towards the opening of the bony cavity (fig. 25).

Retrograde cavity preparation

A small round bur latch grip R1/2 or R1/4 is used to prepare a single surface cavity. Care must be exercised to avoid cutting too far palatally; otherwise, perforation of the root might result and the palatal margins will be weakened (fig. 26). If the root is short, owing to resorption, a minimal bevel is indicated and a Class 2 cavity with a retentive slot may be cut (fig. 27).

Retrograde filling

Before the retrograde filling is inserted, the bony cavity should be packed to prevent accidental loss of excess

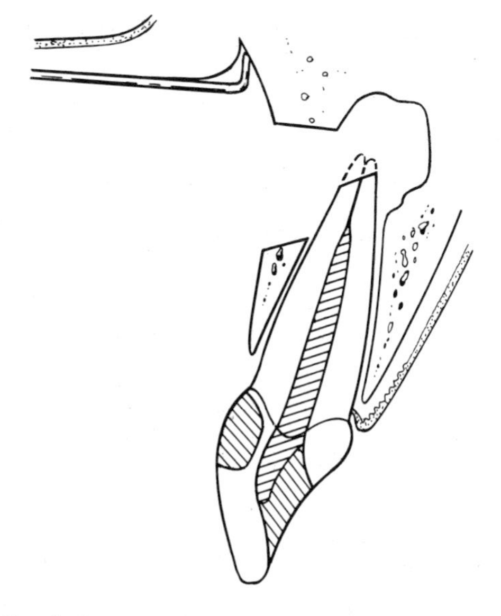

Fig. 25 Bevelled apex: sagittal view.

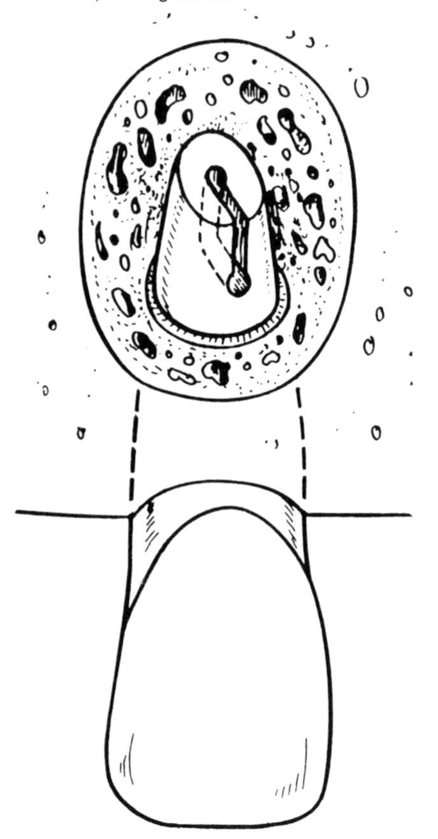

Fig. 26 Completed apicectomy with retrograde amalgam in place.

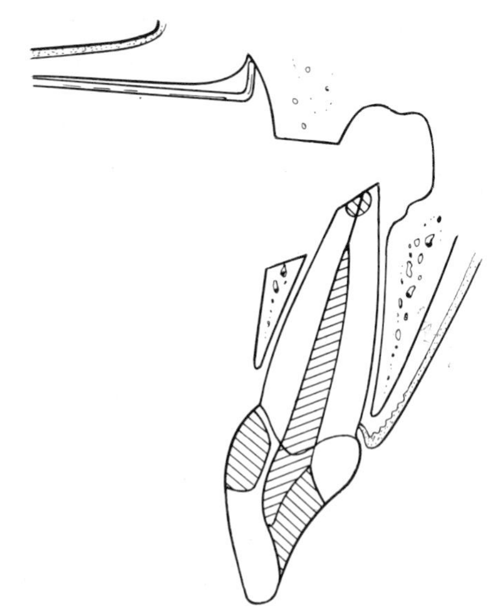

Fig. 27 A Class 2 apical cavity preparation is used when access is limited.

Fig. 28 K-G Retrofilling amalgam carrier (designed by Drs Henry Kahn and Harold Gerstein).

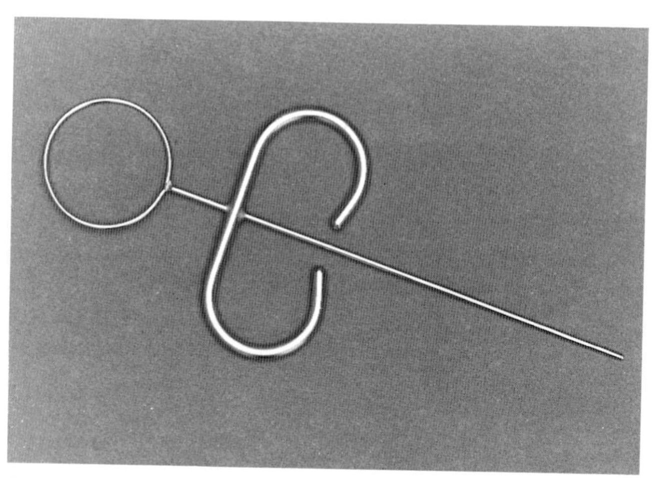

Fig. 29 Retrograde amalgam gun (designed by Professor I. E. Barnes).

filling material around the root. Bone wax or ribbon gauze may be used to isolate the root tip. If gauze is used, it may be wetted with local anaesthetic solution or saline once it is in place, then dabbed dry with a cotton wool pledget.[3,8] Any excess filling material is more easily retained by the damp gauze. A zinc-free amalgam is generally advised and, after mixing, this is packed into the retrograde cavity at the root apex using the Hill amalgam carrier, the K-G Retrofilling amalgam carrier (fig. 28), or the 'Retrograde amalgam gun' designed by Professor I. E. Barnes (fig. 29), then condensed.

Amalgam is not the ideal retrograde filling material because of the problems of leakage.[14] However, it does have the advantage that it is easy to handle, is radiopaque and is well tolerated by the tissues. Other materials are being tried, but the necessary research is not complete.

Replacement of flap and suturing

Once the retrograde filling has been completed, the packing around the root is removed and any further debridement is carried out. At this point, infiltration of a little more local anaesthetic solution into the flap and surrounding gingivae may be required, as the effects of analgesia have sometimes lessened. This will make suturing more comfortable for the patient. Interrupted sutures or circumferential sutures may be used, according to the circumstances and the operator's preferences.[2,8] The sutures should be removed after 5–7 days. Antibiotic cover is only necessary if the medical history requires it, but each case should be judged on its merits.

Post-operative discomfort generally lasts for a few days only and the patient should be warned to expect it. Swelling does occur and crushed ice put into a small plastic bag with a small towel round it, should be placed gently against the area at intervals to alleviate discomfort. Analgesics should be prescribed. Warm salt water mouth-baths or Corsodyl solution (chlorhexidine) will aid healing and keep the area clean. A radiograph should be taken at completion of the operation and compared with one taken 6 months later, to monitor healing.

References

1 Ingle J I, Beveridge E E, Cummings R R, *et al.* Endodontic surgery. *In* Ingle J I, Beveridge E E (eds). *Endodontics.* 2nd ed. pp 594–684. Philadelphia: Lea and Febiger, 1976.

2 Arens D E, Adams D W, DeCastro R. *Endodontic surgery.* pp 2–55, 129–139. New York: Harper and Row, 1981.

3 Gutmann J L. Principles of endodontic surgery for the general practitioner. *Dent Clin North Am* 1984; **28:** 895–908.

4 Luebke R G. Surgical endodontics. *Dent Clin North Am* 1974; **18:** 379–391.

5 Barnes I E. *Surgical endodontics. A colour manual.* pp 9–18, 53–61. Lancaster: MTP Press Ltd, 1984.

6 Arens D E. Surgical endodontics. *In* Cohen S, Burns R C (eds). *Pathways of the pulp.* 3rd ed. pp 613–642. St. Louis: C V Mosby, 1984.

7 Weine F S, Gerstein H. Periapical surgery. *In* Weine F S (ed). *Endodontic therapy.* 3rd ed. pp 408–476. St. Louis: C V Mosby, 1982.

8 Harty F J. Endodontics in clinical practice. 2nd ed. pp 173–174, 192–194, 197. Bristol: John Wright and Sons, 1982.

9 Harris M. The general and systemic aspects of endodontics. *In* Harty F J. *Endodontics in clinical practice.* 2nd ed. pp 7–21, Bristol: John Wright and Sons, 1982.

10 Guralnick W. Odontogenic infections. *Br Dent J* 1984; **156:** 440–447.

11 Harty F J, Parkins B J, Wengraf A M. Success rate in root canal therapy: A retrospective study of conventional cases. *Br Dent J* 1970; **128:** 65–70.

12 Webber R T. Apexogenesis versus apexification. *Dent Clin North Am* 1984; **28:** 669–697.

13 Webber R T. Traumatic injuries and the expanded role of calcium hydroxide. *In* Gerstein H (ed). *Techniques in clinical endodontics.* pp 185–188. Philadelphia: W B Saunders, 1983.

14 Szeremata-Browar T L, Vancura J E, Zaki A E. A comparison of the sealing properties of different retrograde techniques: An autoradiographic study. *Oral Surg* 1984; **59:** 82–87.

11

Endodontic Problems

Endodontics is a skill which requires the use of delicate instruments in confined spaces. Inevitably, problems will occur, but many of these are avoidable providing the operator exercises care and patience. A few tips on how to overcome some of these problems will be given in this chapter. Should the reader require a more wide ranging and detailed account he is referred to Gutmann *et al.*, *Problem solving in endodontics.*[1]

Access and rubber dam

Access

It is important to have good visual access and sufficient space to allow direct line access into the apical third of the root canal. A useful way of assessing a patient for molar endodontics is that the operator should be able to place two fingers between the maxillary and mandibular incisors. If this is not possible owing to a small mouth or limited opening, then it is unwise to commence root canal therapy.

Assessing access for posterior surgical endodontics may be done by retracting the lip at the corner of the mouth with a finger; the surgical area should be directly visible.

The general guidelines for access cavities have already been discussed in Chapter 5. However, there are occasions when these should be adapted to suit a particular case. When access to the back of the mouth is difficult, the mesial marginal ridge of the tooth may be reduced. Inadequate access will lead to poor treatment and, unless the endodontic treatment is successful, further restoration of the tooth is irrelevant.

Before cutting the access cavity, the extent and type of final restoration should be borne in mind. If an anterior tooth requires a crown following root treatment, the access cavity could be cut on the labial surface (fig. 1). In posterior teeth it may be advantageous to reduce the walls, if either they are already weakened or there is a crown or root fracture.

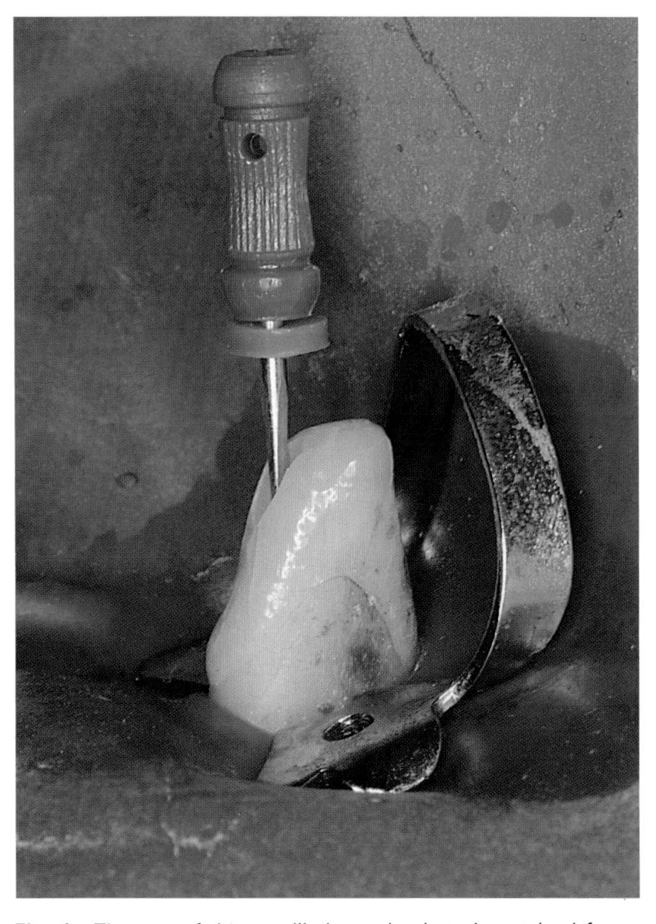

Fig. 1 The root of the mandibular canine is to be retained for an overdenture following root canal treatment. The cavity has been cut buccally to give better access.

Rubber dam

Rubber dam is necessary for medicolegal and practical purposes for root canal treatment. Three situations where its application presents difficulties are described.

The broken down tooth

The broken down tooth may be tackled in a variety of ways. Many teeth with large deficiencies may have rubber dam applied, providing the right clamp is used; the author recommends an ivory W8a or 8a (fig. 2). The clamp is placed directly on to the tooth, so that there is a four-point

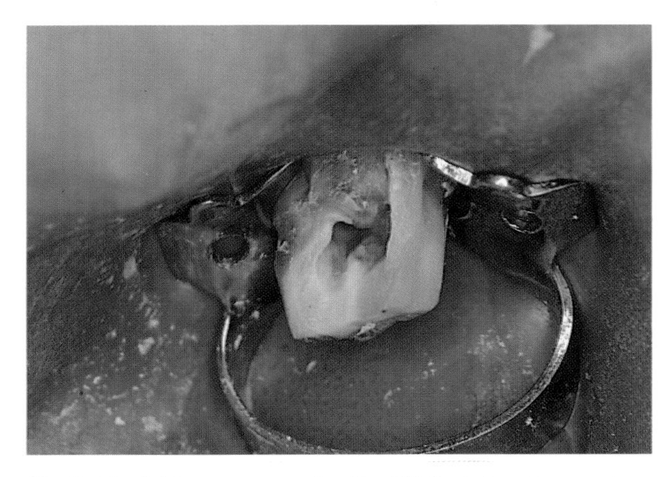

Fig. 2 The bridge was removed as the 6|(16) abutment was carious. The tooth was broken down but a no. W8 clamp fitted allowing good isolation and access for root treatment.

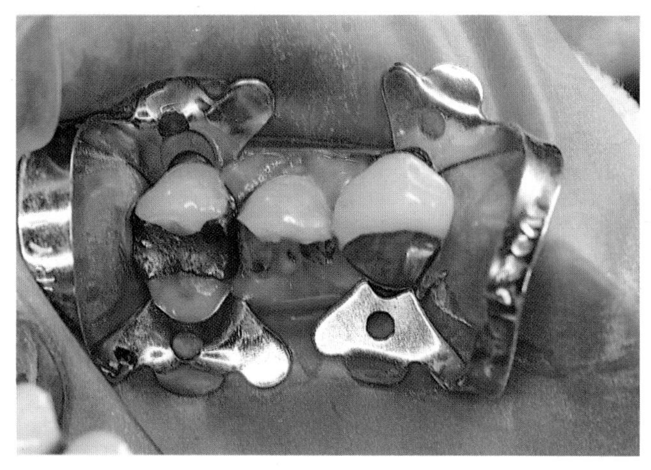

Fig. 3 The 4|(14) is severely broken down, with the palatal wall missing. A no. 1 ivory clamp is placed on the 5|(15) and a no. 0 on the 3|(13). The dam is punched for the 5|(15) and 3|(13) and a slit is cut between the two holes.

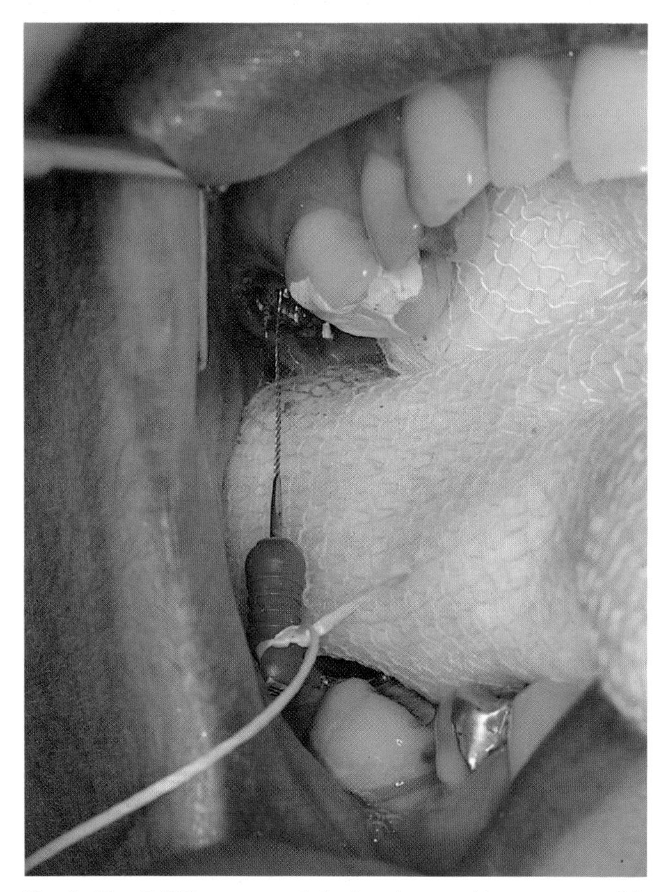

Fig. 4 The 7|(17) was severely broken down and it was not possible to place a rubber dam. Each instrument had floss tied to the handle and a gauze was used for additional protection of the patient's oropharynx.

contact between the jaws of the clamp and the root. Once the clamp is in position, it is checked for stability by pressing with a finger on the bow. As a general rule, before placement of the clamp a length of floss is tied around the bow and the trailing edge allowed to hang out of the mouth. If the clamp were to become dislodged while placing the dam, it can easily and quickly be removed from the patient's mouth. The second method is to clamp the tooth distal to the one broken down, cut a slit between holes punched in the dam, and, if necessary, reverse a clamp on the mesial tooth (fig. 3).

It is feasible to build up the tooth using glass ionomer (for example ketac silver*) or amalgam, although the latter would usually require pins. In the author's opinion, it is inadvisable to place pins and a temporary restoration so that the tooth may be root-filled. The placement of pins will further weaken the tooth and risk a dentine fracture. The method is tedious for both the patient and operator and will result in a reduced access visibility. Cementing an orthodontic band around the tooth using zinc phosphate is a useful option.

On occasion a clamp may be fitted on to a broken down tooth, but only if the gingival tissue encroaching on to the margin is first removed with electrosurgery or a surgical blade.

Finally, on the very rare occasion it is not possible to place rubber dam, the tooth may be root-treated, providing each hand file has floss tied around the handle or a safety device is used. The excess saliva is controlled with cotton wool rolls and a salivary ejector (fig. 4).

Bridges

Bridges do not present a problem with the application of rubber dam. A suitable winged clamp is fitted onto the abutment tooth and the dam stretched over the clamp. If there are any small gaps, these can be sealed with Cavit.

If a bridge is decemented, it must be removed. This should be carried out before any attempt is made to root-treat one of the abutment teeth.

Anterior teeth

Anterior teeth are simple to isolate, giving good access to the cingula without the use of clamps. An example is given in figure 5, where the |1 (21) was to be root-treated. Holes are punched for the anterior six incisors, and the dam placed and retained using wedges cut from a corner of the dam. A slit has been cut in the dam from 1|(11) to |2 (22) to give access to the |1 (21). If better visual access is required, a no. 1 ivory clamp may be placed over the rubber dam on to a premolar, on one or both sides of the mouth.

Locating and negotiating fine canals

Many root canals, particularly in the elderly patient, are difficult to locate. The pulp chambers may be sclerosed or contain large pulp stones and the root canals may be so fine that even when located they are difficult to negotiate.

Dentine deposition occurs as a response to any moderate injury to the pulp, in particular luxation injuries. Initially, the pulp chamber reduces in size, followed by a gradual narrowing of the root canals. The incidence of pulpal necrosis following canal obliteration is not high and so does not warrant intervention by elective root canal treatment.[2]

Radiographs of teeth showing apparent total canal obliteration are deceptive (fig. 6). Andreason[2] refers to a study in which attempts were made to locate and negotiate root canals which were not visible on the pre-operative radiographs. In 54 incisors with periapical lesions, the root canal was located and treated in all but one of them.

*Details can be found at the end of the chapter.

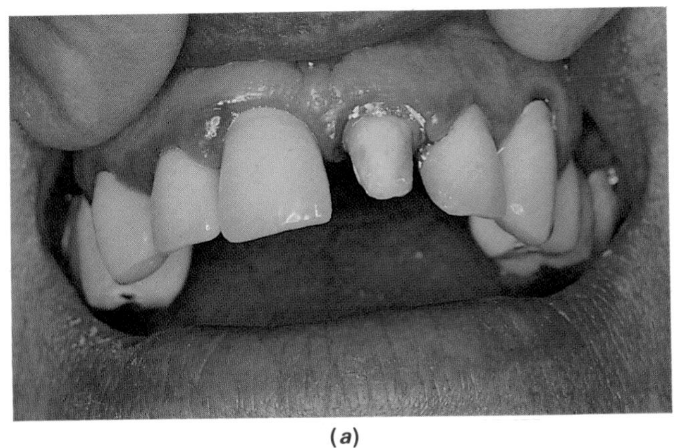

(*a*)

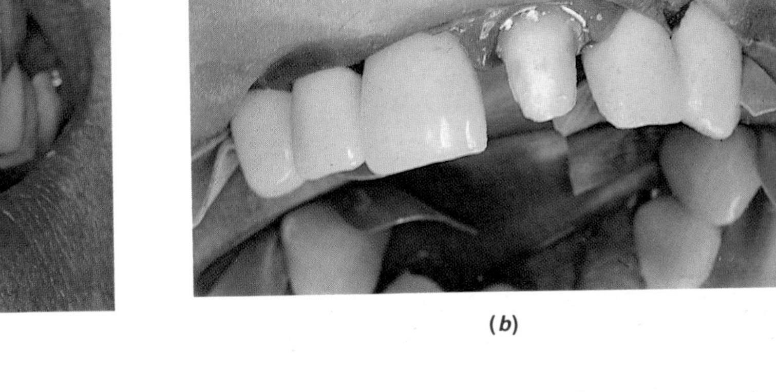

(*b*)

These narrow canals may take time to locate. The pre-operative radiograph contains useful information: the size, curvature and position of the root canal(s) in relation to the pulp chamber are noted. A meticulous search must be made of the floor of the pulp chamber with either a canal probe or an 08 or 10 file. The floor of the pulp chamber is darker than the walls (fig. 7) and the canal entrances are situated, in posterior teeth, at each corner. Fibre-optic light transilluminating the tooth and binocular loops (fig. 8) are also of assistance in difficult cases. If the canal cannot be located, it will be necessary to drill using a small round bur (*see* Chapter 3) in a slow running standard handpiece. A bur hole, approximately 2·0 mm in depth, is then drilled at the expected site of the canal along the main axis of the root. A radiograph is taken with the bur *in situ* and the direction of the bur corrected if further drilling is necessary (fig. 9).

Once the entrance has been located, the next step is to negotiate the canal using a fine instrument. A curve is placed at the tip of an 06 or 08 hand instrument. It is useful to dip the tip of the instrument into a lubricant such as Hibiscrub. The instrument is gradually advanced into the canal using a contra-rotating movement of about one quarter to one half a turn. Force should not be used. The curve in the instrument tip will seek the path of least resistance and allow the instrument to penetrate further into the canal. A push–pull filing motion is used to free coronal obstructions in the canal. The file is removed, copious irrigation used and the procedure repeated until the canal is negotiated to the working length. An 06 instrument will be seen on a diagnostic radiograph.

EDTA paste (ethylenediamine tetra-acetic acid) is not recommended for the initial negotiation of the canal, as it is a chelating agent. The walls of the dentine will be softened, which means a false canal could be cut. EDTA paste is useful in widening the canal walls once the full length has been negotiated.

Ledged or blocked canals

Incorrect technique in preparation can lead to either obstruction of the root canal with dentine debris or the formation of a ledge in the wall of the canal.

In the case of a ledged canal (fig. 10), a curve is placed near the tip of a fine hand instrument, the canal irrigated

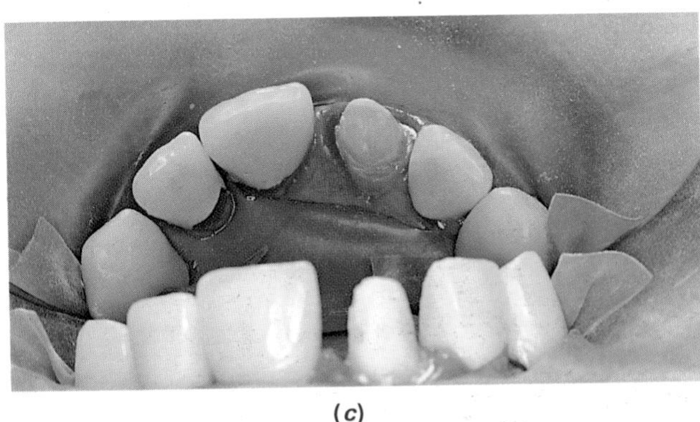

(*c*)

Fig. 5 (*a*) The 1|(11) was to be root-filled. (*b*) The dam had holes punched for the six incisors, a slit was cut to allow the 1|(11) to be bypassed. (*c*) The dam was placed and retained by inserting rubber wedges behind the canines.

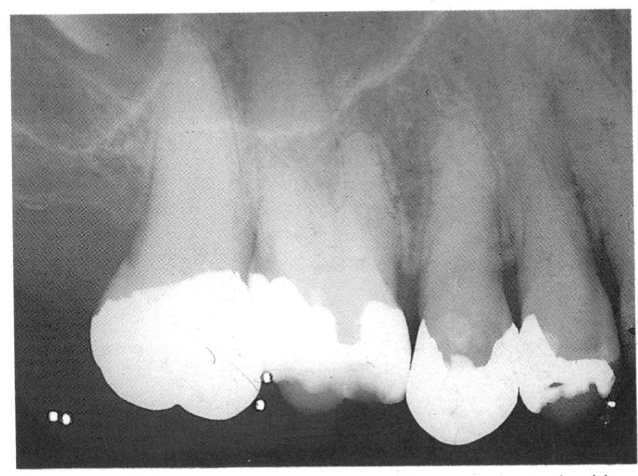

Fig. 6 Radiograph showing a quadrant of four posterior teeth with no root canals apparent.

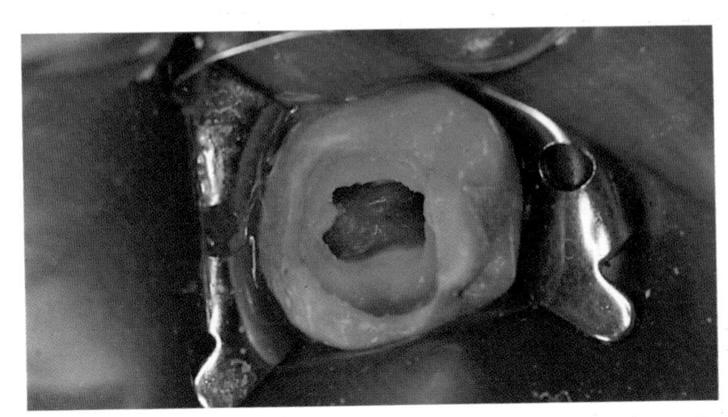

Fig. 7 The floor of the pulp chamber is darker than the dentine walls, as is shown in this molar tooth.

Fig. 8 Binocular loops are particularly useful in endodontics, to help locate the fine entrances to root canals in the floor of the pulp chamber and also to examine the tooth for crown or root fractures.

with sodium hypochlorite, and the instrument inserted into the canal. The instrument tip is directed away from the ledge and gradually advanced with small contra-rotating movements. Once the instrument is beyond the ledge, a push–pull filing motion is used. A lubricant such as EDTA paste is useful to help remove the ledge. This is not a difficult procedure once the ledge has been bypassed (fig. 11).

A canal that has been blocked with dentine debris may well be impossible to negotiate. Copious irrigation, the use of EDTA paste, and a fine instrument may be tried. The danger is of either packing the debris harder into the canal or creating a false canal.

Reroot treatment

A root filling may have to be removed and the tooth retreated for a variety of reasons. The patient may be experiencing symptoms, a periapical radiolucency may be increasing in size, or the coronal restoration may require replacing in a tooth where the root filling is inadequate. Whatever the reason, the first step is to identify the type of filling material that has been used (fig. 12) and to assess the difficulty of the procedure. The method used to remove the previous root filling will depend on the type of material used.

Paste

A soft root-filling material may be removed easily with Hedstroem files and copious irrigation.

Cement

Some cements set hard and have apparently no solvent and, as a result, are almost impossible to remove. The first stage in attempting to remove a cement is to flood the canal entrance with chloroform or xylene and use a canal probe

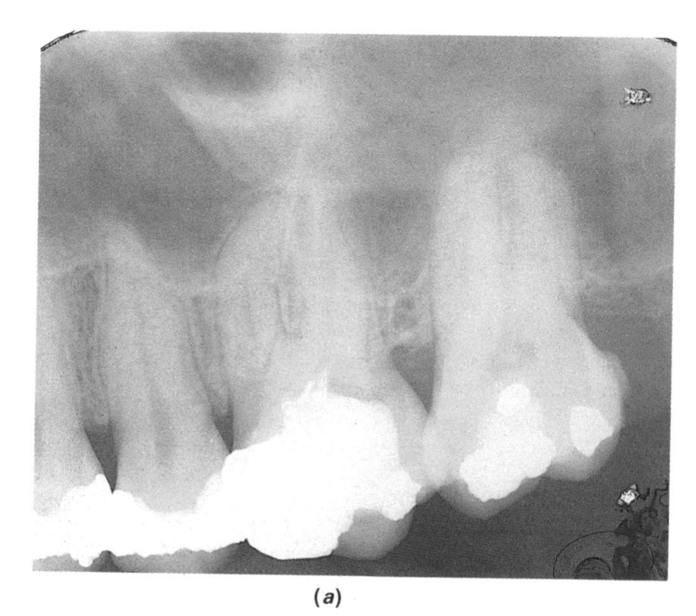

(*a*)

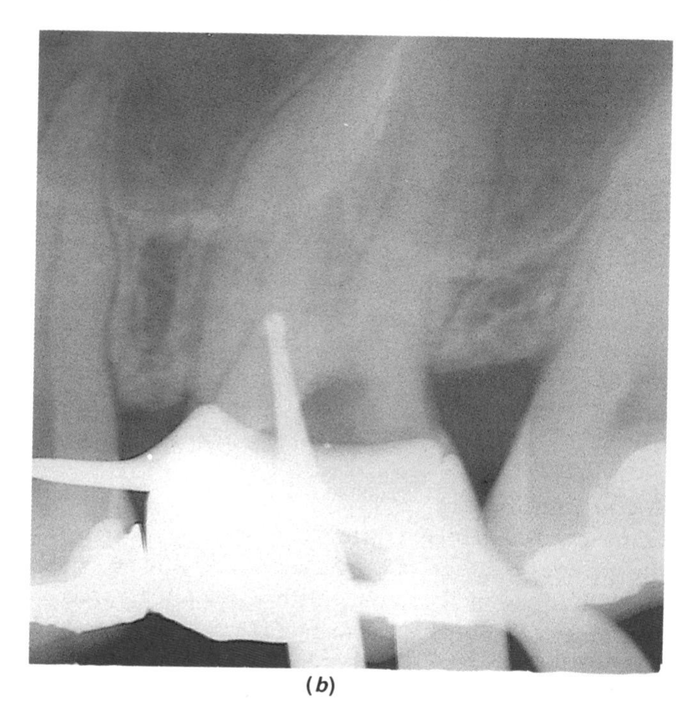

(*b*)

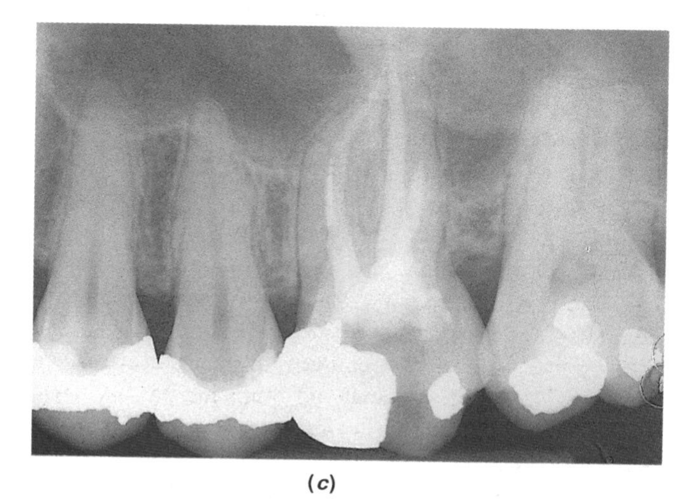

(*c*)

Fig. 9 (*a*) The coronal portion of the mesiobuccal root canal in the |6 (26) was sclerosed. (*b*) When the entrance could not be located, a bur was used to drill along the main axis of the root. (*c*) The canal was located, prepared and filled. Note the distobuccal root, which was sclerosed and not fully negotiable.

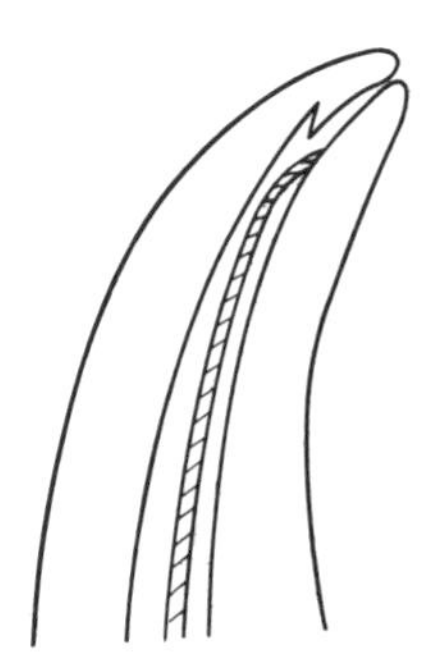

Fig. 10 A root canal containing a ledge may be negotiated by using an instrument which has a curve placed near the tip.

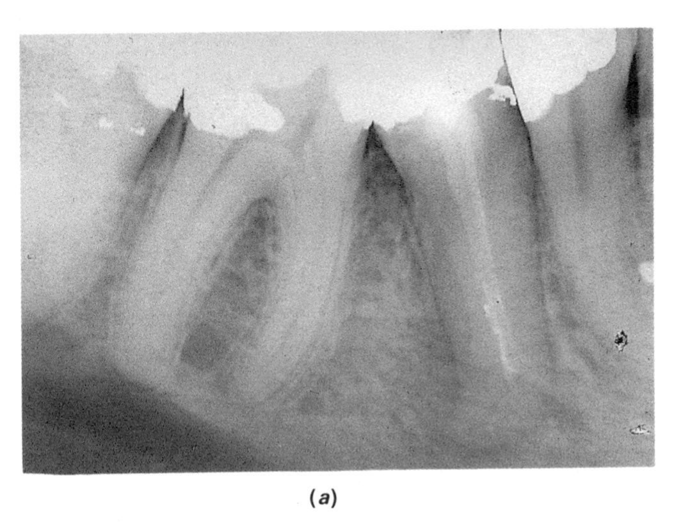

(*a*)

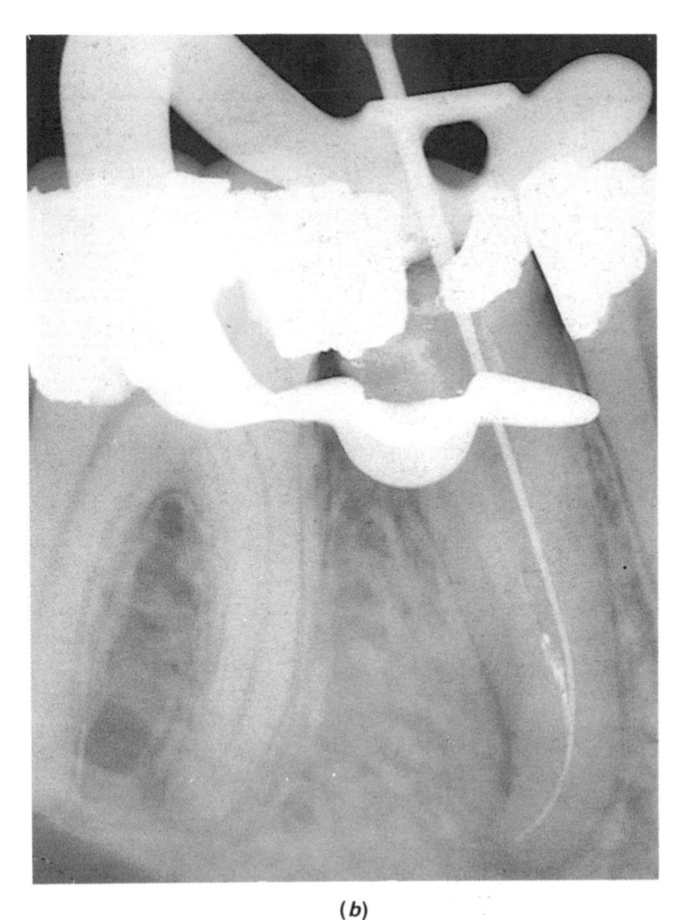

(*b*)

(*c*)

Fig. 11 (*a*) The $\overline{5}|$(45) had been root-filled short, the old root filling was removed and a ledge found in the canal. (*b*) A small curve placed at the tip of the instrument made it possible to bypass the ledge. (*c*) The root canal could then be filled correctly.

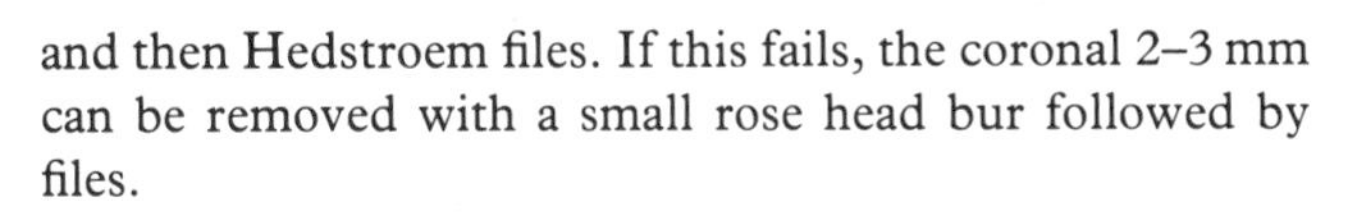

and then Hedstroem files. If this fails, the coronal 2–3 mm can be removed with a small rose head bur followed by files.

Gutta-percha

Gutta-percha is simple to remove. A gap is made between the material and the wall of the canal with a canal probe, and a Hedstroem file is then screwed into the space. A second and, if possible, third file are then inserted and screwed into the mass of the gutta-percha. The handles of the files are grasped, and a steady withdrawal force exerted to remove the gutta-percha point(s) (fig. 13).

Gutta-percha readily dissolves in chloroform and xylene and may then be removed from the canal with Hedstroem files. This technique should not be used when gutta-percha has been extruded beyond the apical foramen; in these cases Hedstroem files are needed to grip and pull back the gutta-percha.

Metal points

The method of removing silver or titanium points is dictated by their position within the root canal (fig. 14). Silver points are easier to remove if there has been leakage of tissue fluids into the canal and corrosion has occurred.

The simplest situation is when the coronal end of the point protrudes far enough into the pulp chamber so that it may be grasped by either Steglitz forceps, narrow-beaked artery forceps or fine pliers.

If the point lies in the root canal below the pulp chamber but in a straight part of the canal, attempts should be made to bypass and either remove the point or

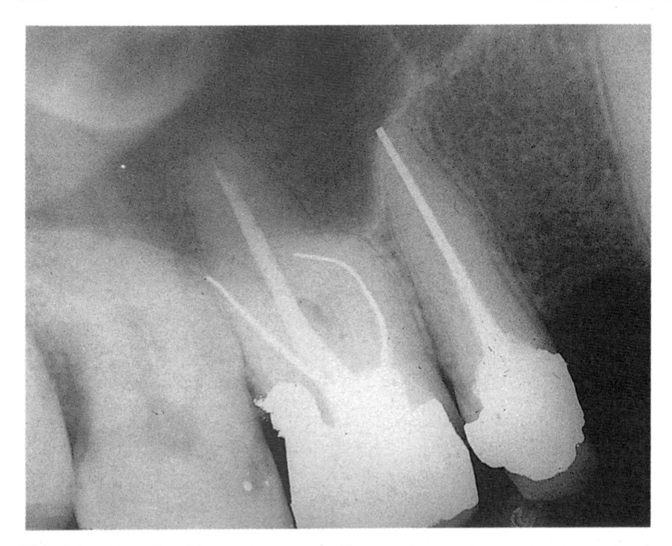

Fig. 12 The 6|(16) has been root-filled using a silver point in each of the buccal canals and gutta-percha in the palatal canal. The operator should be able to distinguish between various root-filling materials from their radiographic appearance.

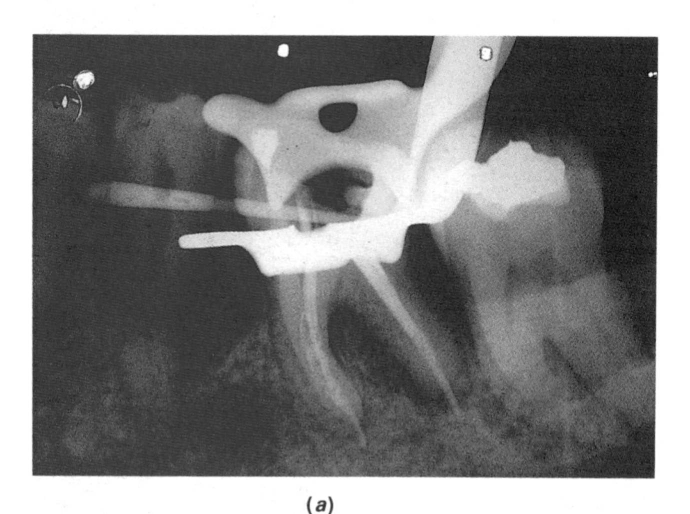

(*a*)

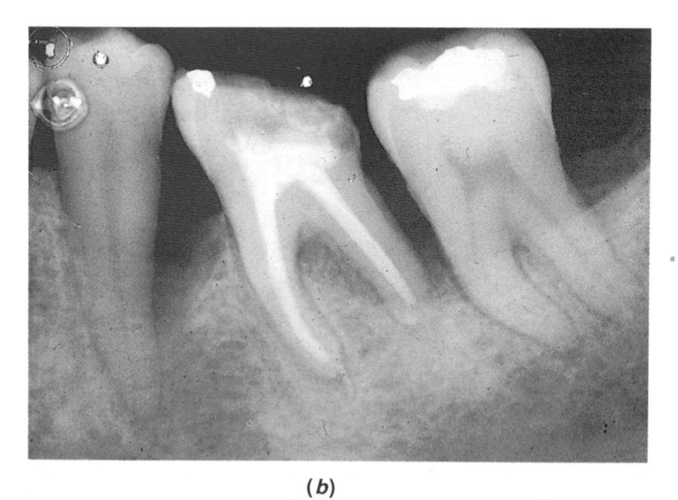

(*b*)

Fig. 13 (*a*) The |6 (36) had been root-filled with gutta-percha extruded through the apices of both roots. (*b*) The root filling was redone removing the gutta-percha with Hedstroem files. Note the excess filling which has been withdrawn from the straight distal canal, but the extruded material remains around the mesial root. It is usually not possible to remove the gutta-percha which has been extruded through the apex in a curved canal.

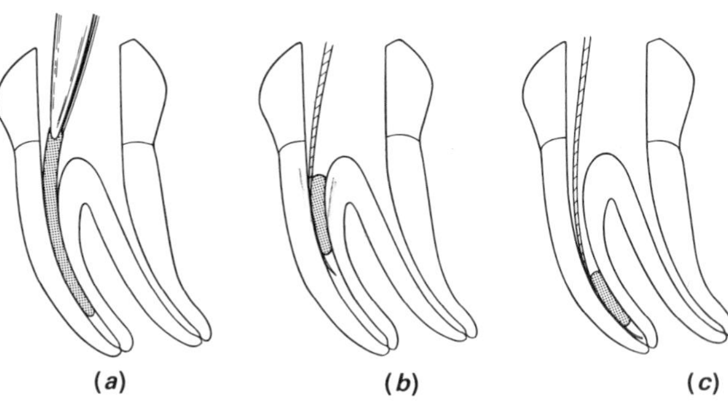

(*a*) (*b*) (*c*)

Fig. 14 Techniques for removing metal points or instruments. (*a*) Protruding into pulp chamber: Steglitz or narrow-beaked artery forceps. (*b*) Straight part of canal: bypass and remove or incorporate into root filling. (*c*) Curved part of canal: bypass and incorporate into root filling.

incorporate it into the root filling. A size 08 or 10 file or reamer is used, and the tip is coated with EDTA paste. If the point can be bypassed, it can frequently be removed with Hedstroem files or using the ultrasonic technique.

It is seldom possible to remove a point which is lodged in the apical third of a curved root canal. Attempts should be made either to bypass the fragment and incorporate it into the root filling, or condense gutta-percha vertically up to the obstruction, with a view to apical surgery should signs of failure occur.

Fractured instruments

The time required to remove or bypass a fractured instrument from the root canal far outweighs the simple precautions that should be taken routinely to prevent such an occurrence. The simple rules which will prevent instrument fracture are:

(1) Discard damaged instruments.
(2) Never force an instrument in the canal.
(3) Do not miss out sizes.
(4) Do not rotate an instrument more than one quarter or one half turn in a clockwise direction.

The techniques used for removal of a fractured instrument are similar to those for metal points. In addition, the Masserann kit has been specifically designed to extract metal fragments from root canals.

The Masserann kit (fig. 15) consists of a number of trepans with a range of diameters from 1·1 mm to 2·4 mm. The trepans are hollow tubes designed to cut a trough around the metal fragment (fig. 16). The trough usually has to be cut along at least half the length of the fragment before it is sufficiently loosened to allow its extraction. It is recommended that the trepan is operated by hand, using the special handle provided, and not placed in a handpiece. A feeler gauge from the kit is used to judge the size of the trepan required. EDTA paste will help to lubricate and soften the dentine. The kit also contains a Masserann extractor, which is placed over the end of the loosened fragment so that it may be gripped and removed (fig. 17). If the fragment is too large for the extractor, then a size smaller trepan may be forced over the end of the fragment, which is then gripped firmly enough to allow its withdrawal from the canal.

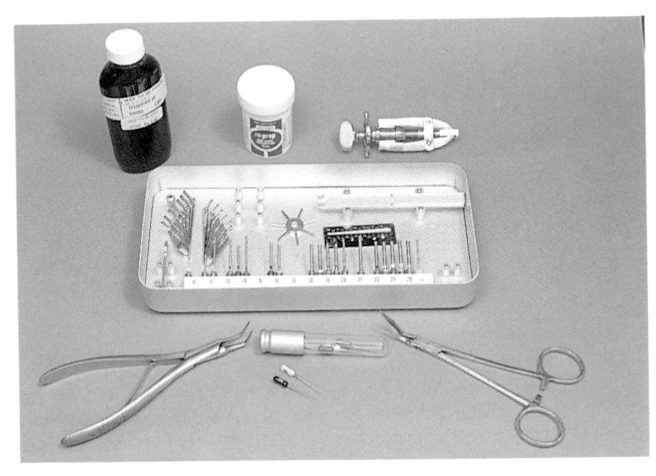

Fig. 15 Instrument and materials used to free obstructed root canals. (*Top row*) chloroform, EDTA paste, Eggler post extractor. (*Middle row*) Masserann Kit. (*Bottom row*) Pin pliers, Hedstroem files, Steglitz forceps.

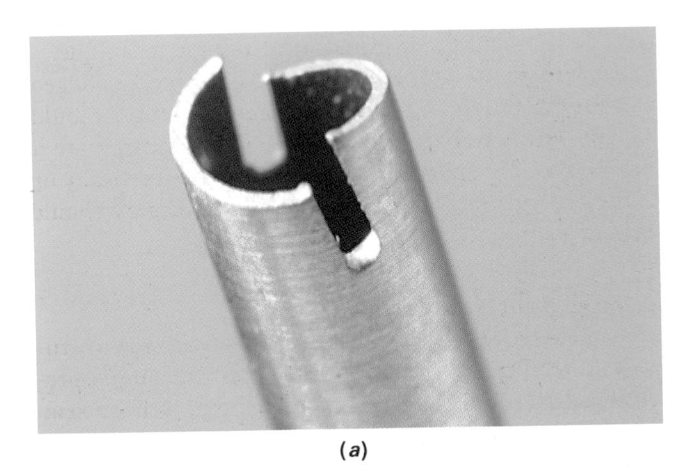

(*a*)

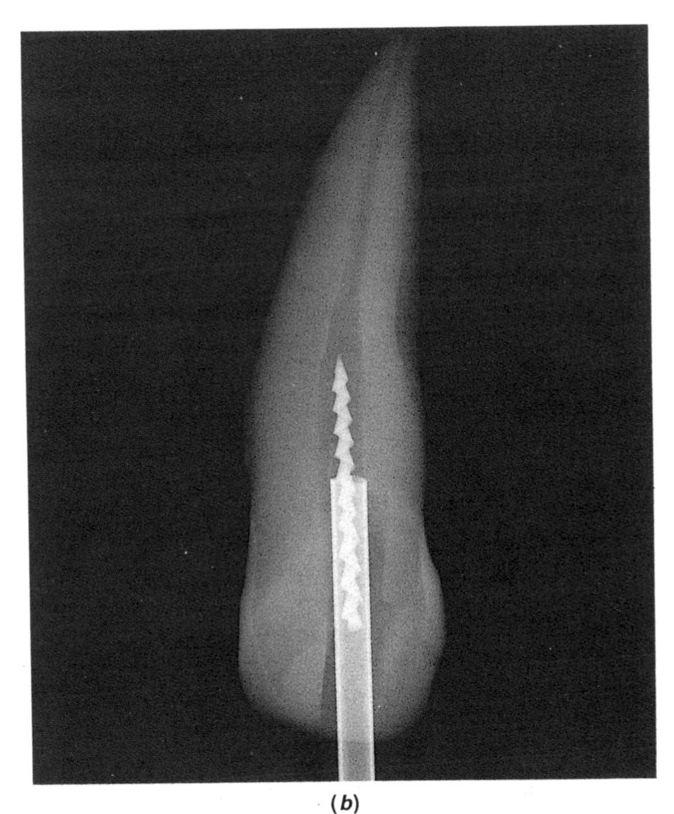

(*b*)

Fig. 16 (*a*) Trepan from the Masserann kit, showing the cutting edge. (*b*) A channel is cut in the dentine around the metal fragment.

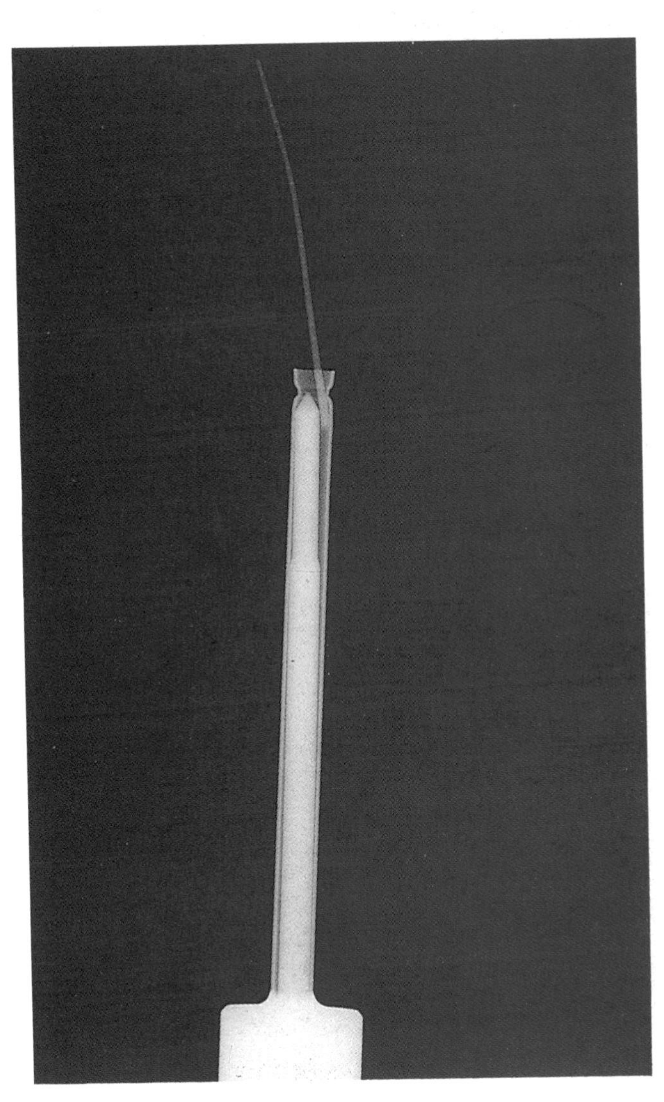

Fig. 17 The Masserann extractor is used to remove fine instruments from the root canal.

A fractured instrument remaining in a canal does not mean that the attempt at root treatment will fail. It has been demonstrated[3,4] that, provided the remainder of the root canal is filled conventionally, the success rate is not significantly affected.

Posts

A post may have to be removed because either the tooth requires (re)root-filling or it has been fractured. The procedure presents problems as there is a danger of fracturing or perforating the root of the tooth. Threaded posts which have fractured can be removed by cutting a groove in the post end and unscrewing. It is possible to extract a smooth-sided post and core using a post extractor (fig. 15). The core must first be shaped so that its sides are parallel and capable of being gripped. The mesial and distal shoulders of the crown preparation must be cut to the same height so there is no torsional force. The post extractor is then placed over the post and the screw tightened onto the core; the feet are then lowered on to the shoulders of the preparation by turning the end knob. Several more turns will ease the post out of the post hole.

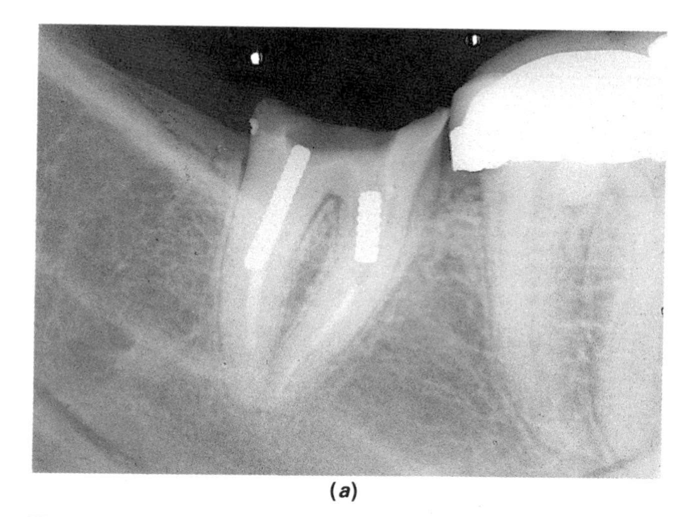

(*a*)

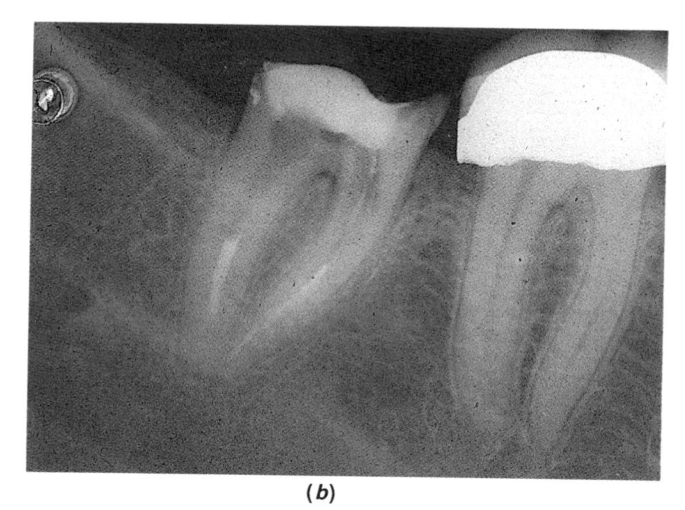

(*b*)

Fig. 18 (*a*) and (*b*) The fractured posts in the $\overline{7}|$(47) were removed by placing the tip of an ultrasonic scaler over the end of the post for 2–3 minutes. The effect of the vibration loosens the cement.

A fractured post lying within the root canal must either be drilled out using a high-speed handpiece, which is a hazardous procedure, or removed with a Masserann kit. An additional alternative is to loosen the post using ultrasound (fig. 18).

References

1 Gutmann J L, Dumsha T C, Lovdahl P E. *Problem solving in endodontics.* Chicago: Year Book Medical Publishers Inc, 1988.

2 Andreason J O. *Traumatic injuries to the teeth.* 2nd ed. pp 183. Copenhagen: Munksgaard, 1981.

3 Crump M C, Natkin E. Relationship of broken root canal instruments to endodontic case prognosis: a clinical investigation. *J Am Dent Assoc* 1970; **80:** 1341–1347.

4 Fox J, Moodnik R N, Greenfield E, Atkinson J S. Filling root canals with files; radiographic evaluation of 304 cases. *NY State Dent J* 1972; **38:** 154–157.

Ketac silver: ESPE, Fabrik Pharmazeutischer Präparate, GbmH & Co KG, D-8031 Seefeld, Oberbayern, W. Germany.

Hibiscrub: ICI Pharmaceuticals division, Macclesfield, Cheshire, England.

EDTA paste: RC Prep. Premier Dental Products Co., Norristown, PA 19401, USA.

Index

Access 8, 87
Access cavity burs 15
Access cavity preparation 32–34
  objectives 33
  pre-treatment radiograph and 32, 34
    stages 34
  root canal morphology and 32–33
  rubber dam and 35
Accessory canals
  morphology 29–32
  primary dentition 69
AH26 18
Amalgam 54, 84
Amputation, root 80–81
Antibiotic prophylaxis 3, 4, 75
Antibiotic therapy 26, 27
Anticurvature filing 38, 40
Apex, open 80
Apexification 55, 80
Apexogenesis 80
Apical periodontitis, acute 23, 24, 25, 27
  over-instrumentation and 26
Apical root resorption, idiopathic 61
Apicectomy 1, 27, 77
  indications 77
  technique 82–85
    analgesia 82
    apex location 82–83
    curettage 83
    flap replacement 84–85
    reflection of flap 82
    retrograde cavity preparation 83
    retrograde filling 83–84
    root resection 83
    suturing 85
Artery forceps 11
Autoclave 11,17
Automated canal preparation devices 39

Balanced force technique 38
Barbed broach 14
Bead steriliser 17, 46
Beechwood creosote 72
Benzalkonium 17
Biocalex 54
Binocular loops 89
Blood dyscrasias 75
Briault probe 11
Burs 15–16, 20
  access cavity cutting 15, 25, 34
  canal location 15
  canal preparation 15–16
  sterilisation 17

Calcinosis 60
Calcium hydroxide 9, 15, 44, 45–46,
    53–55
  apexification and 55, 80
  apexogenesis and 80
  application technique 54
  horizontal fractures and 55
  inter-appointment dressings 55, 58
    long-term 55
  mode of action 53
  perforation closure and 55, 58, 80
  permanent dentition in children
      and 72
  presentations 53–54
  primary dentition and 69, 70–71
  pulp capping
    in children 69, 70
    direct 54–55, 70
    indirect 54, 69, 70
  resorption and 55, 58, 59
  sealers 19
  'weeping' canal 55
Canal master 39
Canal probe 90, 91
Canine, canal morphology
  mandibular 31
  maxillary 30
Cavit 44, 54, 88
Cellulitis 25
Cement filling removal 90–91
Cervical root resorption 61–62
Chemical sterilisation 11, 17
Children 69–73
  aims of treatment 69
  immature tooth pulpotomy 55
  permanent dentition 72–73
    damage from primary tooth
        pathology 69
    first premolar treatment 72
    non-vital pulp and 72–73
    vital pulp and 72
  primary dentition *see* Primary
      dentition
  suitability for treatment 69
Chlorhexidine mouthwash 3, 85
Chloropercha 18
Cold pulp testing 7, 23
Contra-indications to treatment 8–9
Corticosteroid intracanal medication 19,
    24, 44
  emergency treatment and 24, 25
Corticosteroid therapy 3, 75
Cotton wool pledgets 11

Cotton wool rolls 11
Cracked tooth syndrome 23, 25–26
  symptoms 26
Cresatin dressing 24, 25
Crown-down pressureless technique 37
Curved canals 9
  anticurvature filing and 38
  balanced force technique and 38
  flexible instruments and 37, 38–39
  modified instrument tips and 37, 38
  stepback technique and 37
Cuspal height reduction 35
Cutting instruments,
    power-assisted 14–15

Dentine debris impaction 37
Dentine deposition 80, 88
Diabetes mellitus 75
Diagnosis
  case history and 3
  emergencies and 23
  medical history and 3, 4
  pain characteristics and 3, 5
  in primary teeth 69
  pulp testing and 5–6
  radiography and 5
  test cavity cutting 7
  tests 5–7
Diaket 18
Diaket A 19
Diamond burs 15
Disinfection, chemical 17
Double flared technique 37
Dry heat sterilisers 17

EDTA paste 89, 90, 92
Elbow formation 37, 42
Elective root treatment 7, 8
Electronic apex locator 16, 40
  guidelines 40
Electronic pulp tester 23, 24, 55
Emergency treatment 23–27
  acute periapical abscess and 23, 24
  cracked tooth syndrome and 23
  diagnostic aids 23
  history taking 23
  medical history and 23
  post-treatment pain 23, 27
  pulpitis and 23
  treatment-associated pain 23, 26
Endocarditis, infective 3, 8
Endolocking tweezers 11
Endomethasone 19

Eugenol-containing sealers 18
Examination, clinical 5
Excavator 11
Exposure pulpal, pain and 8, 23, 26
Eye protectors 12

Failed root filling retreatment *see* Reroot treatment
Fibre-optic light transillumination 7
 cracked tooth syndrome 26
 fine canal location 89
Files 91, 92
 flexible 38
 sterilisation 17
 storage 17
Filling root canal 1–2, 45–51
 instruments 19–20, 21
 post-treatment pain and 23
 properties of materials 45
 purpose of 45
 sealers and 1
 techniques 45–51
 see also Gutta-percha filling techniques
Fine canals 88–89
Flat plastic 11
Flexible instruments 37, 38–39
Flex-o-file 14
Fluoride varnish application 23
Follicular cysts, inflammatory 72
Formocresol pulpal dressing 71, 72
Fracture, root 23, 26, 27
 calcium hydroxide treatment 55
 incomplete 7, 9
 see also Cracked tooth syndrome
Fractured instruments 92–93
 Masserann kit and 92
 prevention 92
 surgical approach and 80
Full mucoperiosteal flap 82
Furcation probe 11

Gates-Glidden bur 15–16, 41, 43
Gaucher's disease 60
Giromatic handpiece 14, 15
Glutaraldehyde 17, 71
Grossman's sealer 18
Gutta-percha
 filling 1–2, 45, 73
  removal 13, 16, 91
 inter-appointment dressing 44
Gutta-percha filling technique 45–46
 injection-moulded thermoplasticised 49, 51
 lateral condensation 19, 20, 51
  cold 46
  warm 46, 58
 single point 46
 thermatic condensation 48
 vertical condensation 46

Hand instruments 12–14, 20
 colour coding 13
 numbering 13
Handpieces
 power-assisted 14
 ultrasonic 15
Heart disease 75
Heat testing 6–7, 23, 24
Hedstroem files 13, 14, 40, 41, 90, 91, 92
Helifile 14
Heligirofile 15
Helisonic 15
Hemisection, root 80–81
Hepatitis virus 17
High restoration 23, 25, 26, 27
Hill amalgam carrier 84
History taking 3
 emergency treatment and 23
 surgical endodontics and 75
Hollow tube theory 1
Hyperparathyroidism 60

Idiopathic root resorption 60–62
 apical 61
 cervical 61–62
Immature teeth *see* Children
Incision to establish drainage 75
Incisor canal morphology
 mandibular 31
 maxillary 30
Indications for treatment 7–8
Inflammatory root resorption 57, 59
Inhalation sedation 25
Initial tooth preparation 33
Instrumentation, pain following 23, 25 26
Instruments 11–20
 modified tips 37, 38
 pack 11, 20
 storage 17–18
 surgical endodontics 81
Inter-appointment dressing 44, 55
 long-term 55
 with 'weeping' canal 55
Intermediate restorative material (IRM) 44, 54
Internal resorptive lesions 51
Intracanal medication 43–44
Intraligamental injection 25
Intrapulpal analgesia 25
Irrigation 39, 41, 42
 endosonic 42
Irrigation needle 39

K-flex files 14
K-G Retro-filling amalgam carrier 84
Kri liquid/paste 72
K-type files 13, 41, 43

Latch grip bur 82, 83
Lateral canals 1
 morphology 29–32
 perio-endo lesions and 62, 64
 post-periodontal treatment pain and 26
Lateral periodontal abscess 25
Ledged canals 37, 41, 89–90
Leubke-Ochsenbein flap 81
Local anaesthesia 7, 24–25

Magnetostrictive endosonic unit 42
Masserann kit 92, 93
Maxillary sinus disorders 23
McSpadden compactor 48
Measuring devices 16–17
Medical history 3, 8, 23
 surgical endodontics and 75
Medicated sealers 18–19
Messing gun 54
Metal points removal 91–92
Metal ruler 11, 16
Mirror, mouth 11
Mobility testing 5, 25
Modified instrument tips 37, 38
Moist heat sterilisation 17
Molars
 first permanent 72
 mandibular, root canal morphology
  first 31
  second 32
  third 32
 maxillary, root canal morphology
  first 30–31
  second 31
  third 31
Motivation, patient 8
Myofascial pain dysfunction (MPD) syndrome 23

N2 Universal 19
Nogenol 18
Non-eugenol sealers 18
Non-functional teeth 9
Non-vital teeth 7

Obtura delivery system 49
Occlusion 23, 24, 26, 27
Oral hygiene 8
Overdenture 7
Overfilling 23, 27

Pacemakers 6
Paget's disease 60
Palatal grooves 9
Palpation 7
Paper points 11
Paraformaldehyde 1, 19, 44
Pathfinder 39
Percussion 5, 23, 24
Perforation
 calcium hydroxide treatment 55
 curved root preparation and 37
 internal root resorption and 58
 surgical repair 80
Periapical abscess, acute 23, 24, 25
Periapical curettage 75, 77
Periapical lesion biopsy 77
Periodontal disease 8, 35
Periodontal probe 23
Periodontal support 9
Periodontal treatment, pain following 23, 26
Perio-endo lesions 62–66
 classification 64–65
 diagnostic aspects 64, 65
 primary endodontic
  draining through periodontal ligament (Class 1) 64
  secondary periodontal involvement (Class 2) 64
 primary periodontal lesion involvement (Class 3) 64–65
  with secondary endodontic lesion (Class 4) 65
 pulpal–periapical tissue interaction and 62, 64
 root canal treatment 66
 root removal and 65–66
Phoenix abscess 23, 26
Piezoelectric endosonic unit 42
Pocket measuring probe 11
Post space 7, 16
Posts removal 93–94
Post-treatment pain 23, 27
Power-assisted instruments 14–15, 20
 cutting instruments 14–15
 handpieces 14
 spiral root canal fillers 15

Premolars, root canal morphology
mandibular 31
maxillary 30
Preparation, root canal 37–44
anticurvature filing and 38
automated devices and 39
balanced force technique 38
burs 15–16
flexible instruments and 37, 38–39
intracanal medication and 43–44
irrigation 39, 41, 42
modified instrument tips and 37, 38
principles 37–39
procedure 40–43
standardised system 37
stepback technique 37
stepdown technique 37–38, 40–42
temporary restorative materials and 44
ultrasonic technique 42–43
working length measurement 16–17, 39–40, 41, 43
Pressure-associated root resorption 60
Primary dentition
internal root resorption following treatment 72
morphological aspects 69
pulpal pathology diagnosis 69
treatment techniques 69–72
calcium hydroxide dressings 69, 70–71
direct pulp capping 70
formocresol dressings 71
glutaraldehyde dressings 71
indirect pulp capping 69–70
modified pulpectomy 71–72
paraformaldehyde paste 71
vital pulpotomy 70
Probes 11
Protective equipment for patient 12
Pulp capping
direct 54–55, 70, 72
indirect 54, 69–70, 72
permanent dentition in children 72
primary dentition 69–70
Pulp testing
electrical 5–6
rubber gloves and 6
thermal 6
Pulpal sclerosis, post-traumatic 8
Pulpitis
chronic, internal resorption and 57
emergency treatment 23–25
initial treatment 23
local analgesia 24–25
pain characteristics 5, 23, 24
reversible/irreversible 23, 24

Radiation therapy 60
Radiography
apicectomy and 85
emergency treatment and 23
equipment 12, 20
manual film processor 12
viewer 12
film holder sterilisation 17
fine canal location and 88, 89
paralleling and 23, 32, 34
film holder 12
pre-operative 32, 34
pulp capping and 55
pulpitis and 5, 24
root end induction and 55
root filling review 46
root length determination and 16, 40, 43
trauma and 8
Reamer 13, 92
flexible 38
power-assisted 14
sterilisation 17
storage 17
Regional analgesia 25
Replacement resorption 60
Reroot treatment 9–10, 90–92
removal of previous root filling
cement 90–91
gutta-percha 91
metal points 91–92
paste 90
Resorbable paste 69
Resorption, root 55, 57–62
aetiology 55, 57, 59
calcium hydroxide treatment 55, 58, 59, 60
external 9, 59–62, 80
idiopathic 60–62
inflammatory 57, 59
pressure-associated 60
replacement 60
surface 59
systemic 60
internal 9, 57–58
with perforation 58
in primary dentition 72
treatment 58, 80
surgical root repair 80
Restorative treatment
pain following 23, 26
pre-root treatment 35
Retrograde amalgam gun 84
Rheumatic fever 24, 75
Rickert's sealer 18
Rispi 15
Roane technique 38
Root canal system morphology 1, 29–34
access cavity preparation and 32–33
apical constriction 29
canal orifices 29
canine
mandibular 31
maxillary 30
incisors
mandibular 31
maxillary 30
lateral/accessory canals 29–32
molars
mandibular 31, 32
maxillary 30–31
premolars
mandibular 31
maxillary 30
primary dentition 69
pulp chamber configurations 29
surgery and 80
Root end induction *see* Apexification
Root length determination *see* Working length measurement
Round bur 15, 25, 34, 89
Rubber dam 12, 20, 25, 26,35
anterior teeth 88
bridges and 88
broken down tooth 87–88
frame/punch disinfection 17
inability to place 35, 88
indications 12
kit 12
method of application 12
ultrasonic technique and 43

Sealapex 19, 54
Sealers 1, 18–19, 20–21
with calcium hydroxide 54
eugenol 18
gutta-percha filling and 45, 46
medicated 18–19
non-eugenol 18
Sedative dressing 23
Semilunar flap 81
Sodium hypochlorite solution 24, 25, 26, 34, 39, 40, 42, 43, 58
Sonic Air 15, 39
Sorter box 17
Spiral fillers 15, 54
Spreaders 19, 20, 46
Standardised canal preparation system 37
Stepback technique 37, 41
Stepdown technique 37–38
apical instrumentation 41–42
radicular access 40–41
Sterile storage box 18
Sterilisation 17, 20
Stops, marking 16, 40, 41, 43, 54
Storage equipment 20
Surface root resorption 59
Surgical endodontics 75–85
access 81
anaesthesia and 75
apicectomy *see* Apicectomy
flap design 81–82
full mucoperiosteal flap 82
Leubke-Ochsenbein flap 81
semilunar flap 81
history taking and 75
incision to establish drainage 75
indications 80
instruments 81
periapical curettage 75, 77
periapical lesion biopsy 77
reroot treatment 77–78
root amputation/hemisection 80–81
time limits for treatment and 78
Systemic disease, root resorption and 60

Temporary coronal seal 54
Temporary restorative materials 44
Test cavity cutting 7
Test handle 16–17
Test tubes, instrument storage 17
Thermal response testing 6–7, 23, 24
Thermocompaction 49, 51
Trays 11–12, 20
Treatment planning 7–10
Treatment principles 2
Treatment-associated pain 23, 26
Tresiolan 23
Triamcinalone paste 44
Trigeminal neuralgia 23
Trioxymethylene 19
Tubliseal 18
Tungsten carbide bur 15
Turner syndrome 60

Ultrafil delivery system 49
Ultrasonic bath 17
Ultrasonic root canal preparation 42–43, 58
irrigation and 42
procedure 43

Ultrasonic root canal preparation—*contd*
  rubber dam and 43
  theoretical aspects 42
Ultrasonic units 15, 39, 42
Underfilling 27
Unifile 14
Unopposed teeth 9

Vascular pain syndromes 23
Vertical condensation 46

Waterproof bib 12
Woodstick 7
Working length measurement 16–17, 20, 39–40, 41, 43
  electronic apex locator and 40
  radiography and 40

Zinc oxide 54
Zinc oxide/eugenol 25, 72
  cement 69, 71, 72
Zipping 13, 37, 41

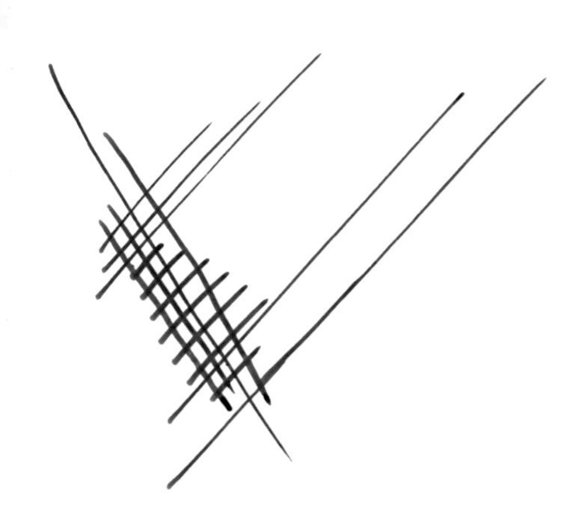